Learn

Eureka Math®
Grade 4
Module 5

Published by Great Minds®.

Copyright © 2018 Great Minds®.

Printed in the U.S.A.

This book may be purchased from the publisher at eureka-math.org.

2 3 4 5 6 7 8 9 10   BAB   25 24 23 22

ISBN 978-1-64054-068-2

G4-M5-L-05.2018

Learn ◆ Practice ◆ Succeed

Eureka Math® student materials for *A Story of Units®* (K–5) are available in the *Learn, Practice, Succeed* trio. This series supports differentiation and remediation while keeping student materials organized and accessible. Educators will find that the *Learn, Practice,* and *Succeed* series also offers coherent—and therefore, more effective—resources for Response to Intervention (RTI), extra practice, and summer learning.

Learn

Eureka Math Learn serves as a student's in-class companion where they show their thinking, share what they know, and watch their knowledge build every day. *Learn* assembles the daily classwork—Application Problems, Exit Tickets, Problem Sets, templates—in an easily stored and navigated volume.

Practice

Each *Eureka Math* lesson begins with a series of energetic, joyous fluency activities, including those found in *Eureka Math Practice.* Students who are fluent in their math facts can master more material more deeply. With *Practice,* students build competence in newly acquired skills and reinforce previous learning in preparation for the next lesson.

Together, *Learn* and *Practice* provide all the print materials students will use for their core math instruction.

Succeed

Eureka Math Succeed enables students to work individually toward mastery. These additional problem sets align lesson by lesson with classroom instruction, making them ideal for use as homework or extra practice. Each problem set is accompanied by a Homework Helper, a set of worked examples that illustrate how to solve similar problems.

Teachers and tutors can use *Succeed* books from prior grade levels as curriculum-consistent tools for filling gaps in foundational knowledge. Students will thrive and progress more quickly as familiar models facilitate connections to their current grade-level content.

Students, families, and educators:

Thank you for being part of the *Eureka Math*® community, where we celebrate the joy, wonder, and thrill of mathematics.

In the *Eureka Math* classroom, new learning is activated through rich experiences and dialogue. The *Learn* book puts in each student's hands the prompts and problem sequences they need to express and consolidate their learning in class.

What is in the Learn book?

Application Problems: Problem solving in a real-world context is a daily part of *Eureka Math*. Students build confidence and perseverance as they apply their knowledge in new and varied situations. The curriculum encourages students to use the RDW process—Read the problem, Draw to make sense of the problem, and Write an equation and a solution. Teachers facilitate as students share their work and explain their solution strategies to one another.

Problem Sets: A carefully sequenced Problem Set provides an in-class opportunity for independent work, with multiple entry points for differentiation. Teachers can use the Preparation and Customization process to select "Must Do" problems for each student. Some students will complete more problems than others; what is important is that all students have a 10-minute period to immediately exercise what they've learned, with light support from their teacher.

Students bring the Problem Set with them to the culminating point of each lesson: the Student Debrief. Here, students reflect with their peers and their teacher, articulating and consolidating what they wondered, noticed, and learned that day.

Exit Tickets: Students show their teacher what they know through their work on the daily Exit Ticket. This check for understanding provides the teacher with valuable real-time evidence of the efficacy of that day's instruction, giving critical insight into where to focus next.

Templates: From time to time, the Application Problem, Problem Set, or other classroom activity requires that students have their own copy of a picture, reusable model, or data set. Each of these templates is provided with the first lesson that requires it.

Where can I learn more about Eureka Math *resources?*

The Great Minds® team is committed to supporting students, families, and educators with an ever-growing library of resources, available at eureka-math.org. The website also offers inspiring stories of success in the *Eureka Math* community. Share your insights and accomplishments with fellow users by becoming a *Eureka Math* Champion.

Best wishes for a year filled with aha moments!

Jill Diniz

Jill Diniz
Director of Mathematics
Great Minds

The Read–Draw–Write Process

The *Eureka Math* curriculum supports students as they problem-solve by using a simple, repeatable process introduced by the teacher. The Read–Draw–Write (RDW) process calls for students to

1. Read the problem.

2. Draw and label.

3. Write an equation.

4. Write a word sentence (statement).

Educators are encouraged to scaffold the process by interjecting questions such as

- What do you see?

- Can you draw something?

- What conclusions can you make from your drawing?

The more students participate in reasoning through problems with this systematic, open approach, the more they internalize the thought process and apply it instinctively for years to come.

Contents

Module 5: Fraction Equivalence, Ordering, and Operations

Topic E: Extending Fraction Equivalence to Fractions Greater Than 1

Topic F: Addition and Subtraction of Fractions by Decomposition

Topic G: Repeated Addition of Fractions as Multiplication

Topic H: Exploring a Fraction Pattern

Use your scissors to cut an index card on the diagonal lines. Prove that you have cut the rectangle into 4 fourths. Include a drawing in your explanation.

_____

_____

_____

_____

Read   **Draw**   **Write**

Lesson 1:   Decompose fractions as a sum of unit fractions using tape diagrams.   1

© 2018 Great Minds®. eureka-math.org

Name _____ Date 2-6-23

1. Draw a number bond, and write the number sentence to match each tape diagram. The first one is done for you.

a.

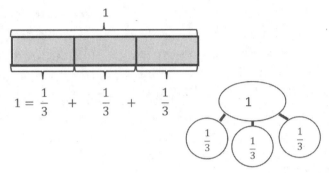

$$1 = \frac{1}{3} + \frac{1}{3} + \frac{1}{3}$$

b.

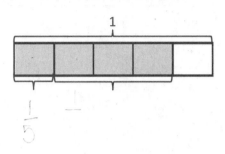

$\frac{1}{5}$

c.

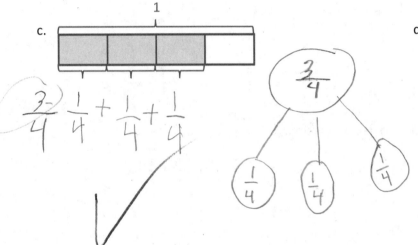

$$\frac{3}{4} = \frac{1}{4} + \frac{1}{4} + \frac{1}{4}$$

d.

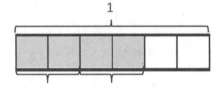

e.

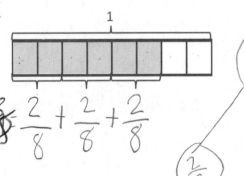

$$\frac{6}{8} = \frac{2}{8} + \frac{2}{8} + \frac{2}{8}$$

f.

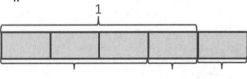

EUREKA MATH

Lesson 1: Decompose fractions as a sum of unit fractions using tape diagrams.

3

© 2018 Great Minds®. eureka-math.org

g.

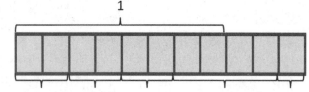

h.

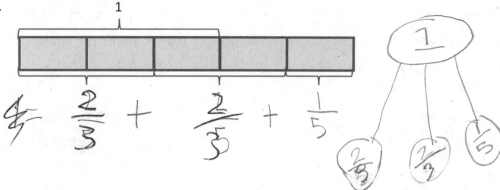

2. Draw and label tape diagrams to model each decomposition.

a. $1 = \frac{1}{6} + \frac{1}{6} + \frac{1}{6} + \frac{1}{6} + \frac{1}{6} + \frac{1}{6}$

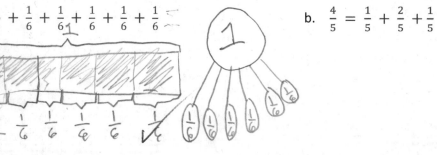

b. $\frac{4}{5} = \frac{1}{5} + \frac{2}{5} + \frac{1}{5}$

c. $\frac{7}{8} = \frac{3}{8} + \frac{3}{8} + \frac{1}{8}$

d. $\frac{11}{8} = \frac{7}{8} + \frac{1}{8} + \frac{3}{8}$

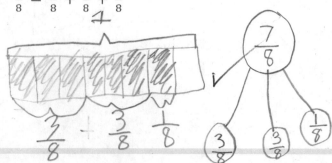

e. $\dfrac{12}{10} = \dfrac{6}{10} + \dfrac{4}{10} + \dfrac{2}{10}$

f. $\dfrac{15}{12} = \dfrac{8}{12} + \dfrac{3}{12} + \dfrac{4}{12}$

g. $1\dfrac{2}{3} = 1 + \dfrac{2}{3}$

h. $1\dfrac{5}{8} = 1 + \dfrac{1}{8} + \dfrac{1}{8} + \dfrac{3}{8}$

Lesson 1:   Decompose fractions as a sum of unit fractions using tape diagrams.

© 2018 Great Minds®. eureka-math.org

5

Name _____   Date _____

1. Complete the number bond, and write the number sentence to match the tape diagram.

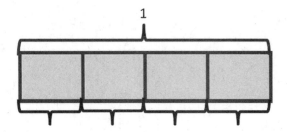   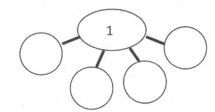

2. Draw and label tape diagrams to model each number sentence.

a. $1 = \frac{1}{5} + \frac{1}{5} + \frac{1}{5} + \frac{1}{5} + \frac{1}{5}$

b. $\frac{5}{6} = \frac{2}{6} + \frac{2}{6} + \frac{1}{6}$

Mrs. Salcido cut a small birthday cake into 6 equal pieces for 6 children. One child was not hungry, so she gave the birthday boy the extra piece. Draw a tape diagram to show how much cake each of the five children received.

_____

_____

_____

_____

Read   **Draw**   **Write**

Name _____   Date 2-7-23 _____

1.   Step 1: Draw and shade a tape diagram of the given fraction.

Step 2: Record the decomposition as a sum of unit fractions.

Step 3: Record the decomposition of the fraction two more ways.

(The first one has been done for you.)

a. $\frac{5}{8}$

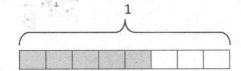

$\frac{5}{8} = \frac{1}{8} + \frac{1}{8} + \frac{1}{8} + \frac{1}{8} + \frac{1}{8}$ $\frac{5}{8} = \frac{2}{8} + \frac{2}{8} + \frac{1}{8}$ $\frac{5}{8} = \frac{2}{8} + \frac{1}{8} + \frac{1}{8} + \frac{1}{8}$

b. $\frac{9}{10}$

$\frac{9}{10} = \frac{1}{9} + \frac{1}{9} + \frac{1}{9} + \frac{1}{9} + \frac{1}{9} + \frac{1}{9} + \frac{1}{9} + \frac{1}{9} + \frac{1}{9} + \frac{1}{9}$

c. $\frac{3}{2}$

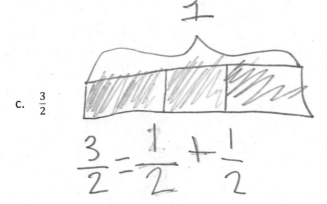

$\frac{3}{2} = \frac{1}{2} + \frac{1}{2}$

EUREKA MATH

Lesson 2:   Decompose fractions as a sum of unit fractions using tape diagrams.

© 2018 Great Minds®. eureka-math.org

11

2. Step 1: Draw and shade a tape diagram of the given fraction.

Step 2: Record the decomposition of the fraction in three different ways using number sentences.

a. $\frac{7}{8}$

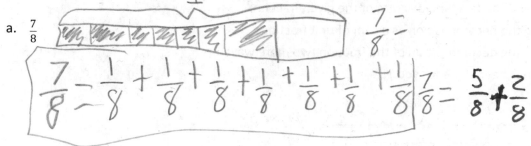

$\frac{7}{8} =$

$\frac{7}{8} = \frac{1}{8} + \frac{1}{8} + \frac{1}{8} + \frac{1}{8} + \frac{1}{8} + \frac{1}{8} + \frac{1}{8}$

$\frac{7}{8} = \frac{5}{8} + \frac{2}{8}$

b. $\frac{5}{3}$

c. $\frac{7}{5}$

d. $1\frac{1}{3}$

Lesson 2:   Decompose fractions as a sum of unit fractions using tape diagrams.

EUREKA
MATH

Mrs. Beach prepared copies for 4 reading groups. She made 6 copies for each group. How many copies did Mrs. Beach make?

    a.   Draw a tape diagram.

    b.   Write both an addition and a multiplication sentence to solve.

Read   **Draw**   **Write**

Lesson 3:   Decompose non-unit fractions and represent them as a whole number times a unit fraction using tape diagrams.

15

c.   What fraction of the copies is needed for 3 groups?   To show that, shade the tape diagram.

Read   **Draw**   **Write**

Lesson 3:   Decompose non-unit fractions and represent them as a whole number times a unit fraction using tape diagrams.

EUREKA
MATH

Name _____   Date 2-8-23

1. Decompose each fraction modeled by a tape diagram as a sum of unit fractions. Write the equivalent multiplication sentence. The first one has been done for you.

a.

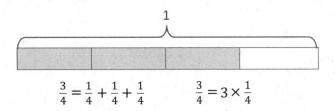

$$\frac{3}{4} = \frac{1}{4} + \frac{1}{4} + \frac{1}{4}$$ $$\frac{3}{4} = 3 \times \frac{1}{4}$$

b.

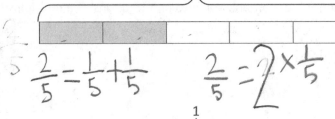

$$\frac{2}{5} = \frac{1}{5} + \frac{1}{5}$$ $$\frac{2}{5} = 2 \times \frac{1}{5}$$

c.

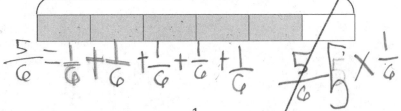

$$\frac{5}{6} = \frac{1}{6} + \frac{1}{6} + \frac{1}{6} + \frac{1}{6} + \frac{1}{6}$$ $$\frac{5}{6} = 5 \times \frac{1}{6}$$

d.

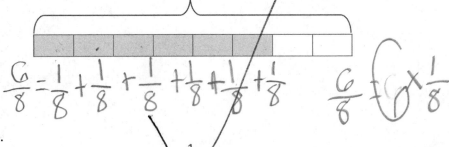

$$\frac{6}{8} = \frac{1}{8} + \frac{1}{8} + \frac{1}{8} + \frac{1}{8} + \frac{1}{8} + \frac{1}{8}$$ $$\frac{6}{8} = 6 \times \frac{1}{8}$$

e.

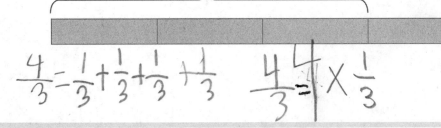

$$\frac{4}{3} = \frac{1}{3} + \frac{1}{3} + \frac{1}{3} + \frac{1}{3}$$ $$\frac{4}{3} = 4 \times \frac{1}{3}$$

EUREKA
MATH

Lesson 3:   Decompose non-unit fractions and represent them as a whole number times a unit fraction using tape diagrams.

© 2018 Great Minds®. eureka-math.org

17

2. Write the following fractions greater than 1 as the sum of two products.

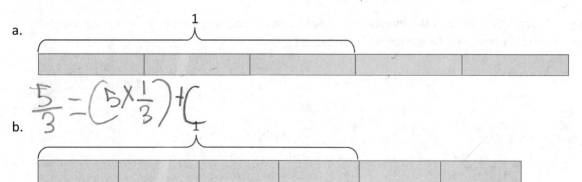

a.

$$\frac{5}{3} = \left(5 \times \frac{1}{3}\right) + C$$

b.

3. Draw a tape diagram, and record the given fraction's decomposition into unit fractions as a multiplication sentence.

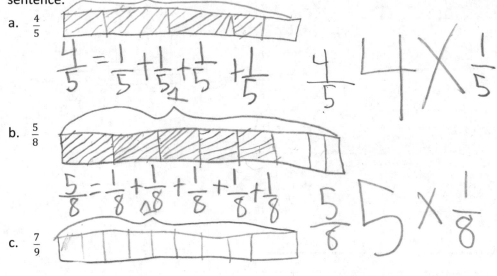

a. $\frac{4}{5}$

$$\frac{4}{5} = \frac{1}{5} + \frac{1}{5} + \frac{1}{5} + \frac{1}{5} \qquad \frac{4}{5} \quad 4 \times \frac{1}{5}$$

b. $\frac{5}{8}$

$$\frac{5}{8} = \frac{1}{8} + \frac{1}{8} + \frac{1}{8} + \frac{1}{8} + \frac{1}{8} \qquad \frac{5}{8} \quad 5 \times \frac{1}{8}$$

c. $\frac{7}{9}$

d. $\frac{7}{4}$

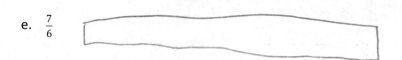

e. $\frac{7}{6}$

Lesson 3:   Decompose non-unit fractions and represent them as a whole number times a unit fraction using tape diagrams.

EUREKA
MATH®

A recipe calls for $\frac{3}{4}$ cup of milk. Saisha only has a $\frac{1}{4}$-cup measuring cup. If she doubles the recipe, how many times will she need to fill the $\frac{1}{4}$ cup with milk? Draw a tape diagram, and record as a multiplication sentence.

_____

_____

_____

_____

Read   **Draw**   **Write**

Lesson 4:   Decompose fractions into sums of smaller unit fractions using tape diagrams.

© 2018 Great Minds®. eureka-math.org

21

Name _____ Date 2-10-23

1. The total length of each tape diagram represents 1. Decompose the shaded unit fractions as the sum of smaller unit fractions in at least two different ways. The first one has been done for you.

a.

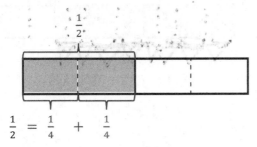

$$\frac{1}{2} = \frac{1}{4} + \frac{1}{4}$$

$$\frac{1}{2} = \frac{1}{8} + \frac{1}{8} + \frac{1}{8} + \frac{1}{8}$$

b.

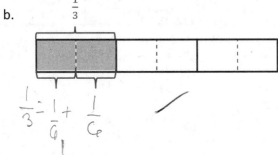

$$\frac{1}{3} = \frac{1}{6} + \frac{1}{6}$$

$$\frac{1}{3} = \frac{1}{12} + \frac{1}{12} + \frac{1}{12} + \frac{1}{12} \checkmark$$

c.

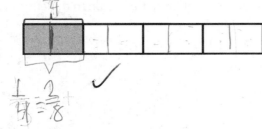

$$\frac{1}{4} = \frac{2}{8} \checkmark$$

$$\frac{1}{4} = \frac{1}{12} + \frac{1}{12} + \frac{1}{12} = \frac{3}{12} \checkmark$$

d.

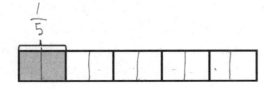

$$\frac{1}{5} = \frac{2}{10} \checkmark$$

$$\frac{1}{5} = \frac{3}{15} \checkmark$$

EUREKA MATH

Lesson 4: Decompose fractions into sums of smaller unit fractions using tape diagrams.

© 2018 Great Minds®. eureka-math.org

23

2. The total length of each tape diagram represents 1. Decompose the shaded fractions as the sum of smaller unit fractions in at least two different ways.

a.

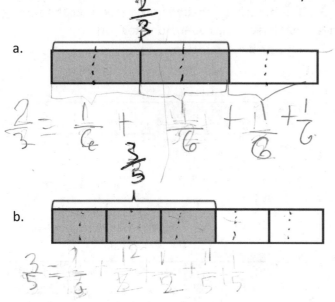

$$\frac{2}{3} = \frac{1}{6} + \frac{1}{6} + \frac{1}{6} + \frac{1}{6}$$

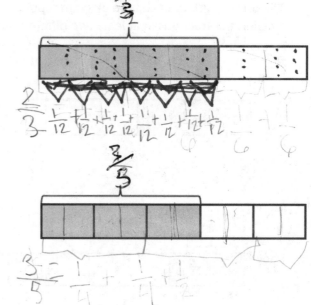

$$\frac{2}{3} = \frac{1}{12} + \frac{1}{12} + \frac{1}{12} + \frac{1}{12} + \frac{1}{12} + \frac{1}{12} + \frac{1}{12} + \frac{1}{12}$$

b.

$$\frac{3}{5} = \frac{1}{5} + \frac{1}{5} + \frac{1}{5}$$

$$\frac{3}{5} = \frac{1}{4} + \frac{1}{4} + \frac{1}{2}$$

3. Draw and label tape diagrams to prove the following statements. The first one has been done for you.

a. $\frac{2}{5} = \frac{4}{10}$

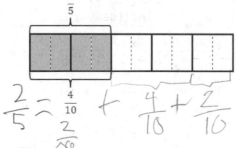

$$\frac{2}{5} = \frac{4}{10} + \frac{4}{10} + \frac{2}{10}$$

$$\frac{2}{6}$$

b. $\frac{2}{6} = \frac{4}{12}$

$$\frac{2}{6} + \frac{4}{12} + \frac{4}{12}$$

Lesson 4: Decompose fractions into sums of smaller unit fractions using tape diagrams.

© 2018 Great Minds®. eureka-math.org

EUREKA
MATH

c. $\frac{3}{4} = \frac{6}{8}$

d. $\frac{3}{4} = \frac{9}{12}$

4. Show that $\frac{1}{2}$ is equivalent to $\frac{4}{8}$ using a tape diagram and a number sentence.

5. Show that $\frac{2}{3}$ is equivalent to $\frac{6}{9}$ using a tape diagram and a number sentence.

6. Show that $\frac{4}{6}$ is equivalent to $\frac{8}{12}$ using a tape diagram and a number sentence.

EUREKA MATH

Lesson 4:　Decompose fractions into sums of smaller unit fractions using tape diagrams.

© 2018 Great Minds®. eureka-math.org

Name _____   Date _____

1. The total length of the tape diagram represents 1. Decompose the shaded unit fraction as the sum of smaller unit fractions in at least two different ways.

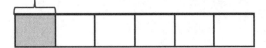 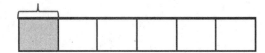

2. Draw a tape diagram to prove the following statement.

$$\frac{2}{3} = \frac{4}{6}$$

Lesson 4:   Decompose fractions into sums of smaller unit fractions using tape diagrams.

© 2018 Great Minds®. eureka-math.org

27

EUREKA
MATH®

A loaf of bread was cut into 6 equal slices. Each of the 6 slices was cut in half to make thinner slices for sandwiches. Mr. Beach used 4 slices. His daughter said, "Wow! You used $\frac{2}{6}$ of the loaf!" His son said, "No. He used $\frac{4}{12}$." Explain who was correct using a tape diagram.

_____

_____

_____

_____

Read   Draw   Write

Name _____   Date _____

1. Draw horizontal lines to decompose each rectangle into the number of rows as indicated. Use the model to give the shaded area as both a sum of unit fractions and as a multiplication sentence.

   a. 2 rows

   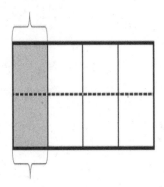

   $$\frac{1}{4} = \frac{2}{}$$

   $$\frac{1}{4} = \frac{1}{8} + \frac{}{} = \frac{}{}$$

   $$\frac{1}{4} = 2 \times \frac{}{} = \frac{}{}$$

   b. 2 rows

   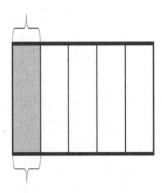

   c. 4 rows

   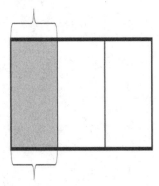

2. Draw area models to show the decompositions represented by the number sentences below. Represent the decomposition as a sum of unit fractions and as a multiplication sentence.

a. $\frac{1}{2} = \frac{3}{6}$

b. $\frac{1}{2} = \frac{4}{8}$

c. $\frac{1}{2} = \frac{5}{10}$

d. $\frac{1}{3} = \frac{2}{6}$

e. $\frac{1}{3} = \frac{4}{12}$

f. $\frac{1}{4} = \frac{3}{12}$

3. Explain why $\frac{1}{12} + \frac{1}{12} + \frac{1}{12}$ is the same as $\frac{1}{4}$.

Lesson 5: Decompose unit fractions using area models to show equivalence.

EUREKA MATH

Name _____   Date _____

1. Draw horizontal lines to decompose each rectangle into the number of rows as indicated. Use the model to give the shaded area as both a sum of unit fractions and as a multiplication sentence.

    a. 2 rows

    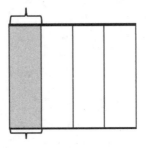

    b. 3 rows

    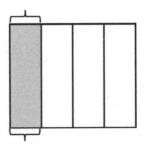

2. Draw an area model to show the decomposition represented by the number sentence below. Represent the decomposition as a sum of unit fractions and as a multiplication sentence.

    $$\frac{3}{5} = \frac{6}{10}$$

Use area models to prove that $\frac{1}{2} = \frac{2}{4} = \frac{4}{8}$, $\frac{1}{2} = \frac{3}{6} = \frac{6}{12}$, and $\frac{1}{2} = \frac{5}{10}$. What conclusion can you make about $\frac{4}{8}$, $\frac{6}{12}$, and $\frac{5}{10}$? Explain.

_____

_____

_____

_____

Read   **Draw**   **Write**

Name _____   Date _____

1. Each rectangle represents 1. Draw horizontal lines to decompose each rectangle into the fractional units as indicated. Use the model to give the shaded area as a sum and as a product of unit fractions. Use parentheses to show the relationship between the number sentences. The first one has been partially done for you.

a. Sixths

$\frac{2}{3}$

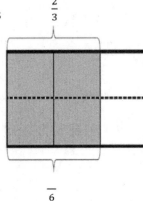

$\frac{2}{3} = \frac{4}{—}$

$\frac{}{3} + \frac{}{3} = \left(\frac{1}{6} + \frac{1}{6}\right) + \left(\frac{1}{6} + \frac{1}{6}\right) = \frac{4}{—}$

$\left(\frac{1}{6} + \frac{1}{6}\right) + \left(\frac{1}{6} + \frac{1}{6}\right) = \left(2 \times —\right) + \left(2 \times —\right) = \frac{4}{—}$

$\frac{2}{3} = 4 \times — = \frac{4}{—}$

$\overline{6}$

b. Tenths

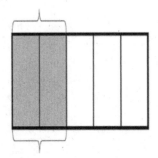

c. Twelfths

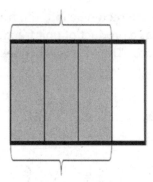

2. Draw area models to show the decompositions represented by the number sentences below. Express each as a sum and product of unit fractions. Use parentheses to show the relationship between the number sentences.

a. $\frac{3}{5} = \frac{6}{10}$

b. $\frac{3}{4} = \frac{6}{8}$

Lesson 6:   Decompose fractions using area models to show equivalence.

EUREKA
MATH

3. Step 1: Draw an area model for a fraction with units of thirds, fourths, or fifths.

Step 2: Shade in more than one fractional unit.

Step 3: Partition the area model again to find an equivalent fraction.

Step 4: Write the equivalent fractions as a number sentence. (If you've written a number sentence like this one already on this Problem Set, start over.)

Lesson 6:   Decompose fractions using area models to show equivalence.

39

© 2018 Great Minds®. eureka-math.org

Name _____   Date _____

1. The rectangle below represents 1. Draw horizontal lines to decompose the rectangle into eighths. Use the model to give the shaded area as a sum and as a product of unit fractions. Use parentheses to show the relationship between the number sentences.

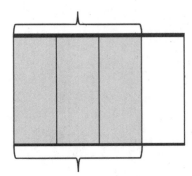

2. Draw an area model to show the decomposition represented by the number sentence below.

$$\frac{4}{5} = \frac{8}{10}$$

Model an equivalent fraction for $\frac{4}{7}$ using an area model.

Read　　　　**Draw**　　　　**Write**

Lesson 7:　　Use the area model and multiplication to show the equivalence of two fractions.

© 2018 Great Minds®. eureka-math.org

43

Name _____ Date 2-20-23

Each rectangle represents 1.

1. The shaded unit fractions have been decomposed into smaller units. Express the equivalent fractions in a number sentence using multiplication. The first one has been done for you.

a.

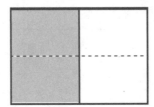

$$\frac{1}{2} = \frac{1 \times 2}{2 \times 2} = \frac{2}{4}$$

b.

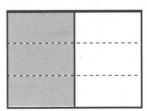

$\frac{2}{2} =$

c.

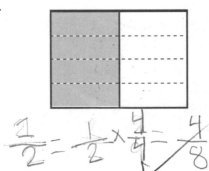

$$\frac{1}{2} = \frac{1}{2} \times \frac{4}{4} = \frac{4}{8}$$

d.

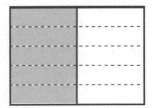

EUREKA MATH

Lesson 7:   Use the area model and multiplication to show the equivalence of two fractions.

© 2018 Great Minds®. eureka-math.org

45

2. Decompose the shaded fractions into smaller units using the area models. Express the equivalent fractions in a number sentence using multiplication.

a.

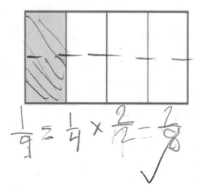

$$\frac{1}{9} = \frac{1}{4} \times \frac{2}{12} = \frac{2}{8}$$ ✓

b.

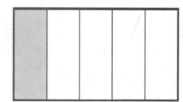

c.

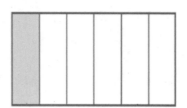

d.

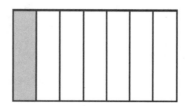

e. What happened to the size of the fractional units when you decomposed the fraction?

iT got smaller ✓

f. What happened to the total number of units in the whole when you decomposed the fraction?

it got more ✓

Lesson 7:   Use the area model and multiplication to show the equivalence of two fractions.

© 2018 Great Minds®. eureka-math.org

EUREKA
MATH

3. Draw three different area models to represent 1 third by shading.
   Decompose the shaded fraction into (a) sixths, (b) ninths, and (c) twelfths.
   Use multiplication to show how each fraction is equivalent to 1 third.

   a.

   b.

   c.

Lesson 7:   Use the area model and multiplication to show the equivalence of two fractions.

© 2018 Great Minds®. eureka-math.org

47

Saisha gives some of her chocolate bar, pictured below, to her younger brother Lucas. He says, "Thanks for $\frac{3}{12}$ of the bar." Saisha responds, "No. I gave you $\frac{1}{4}$ of the bar." Explain why both Lucas and Saisha are correct.

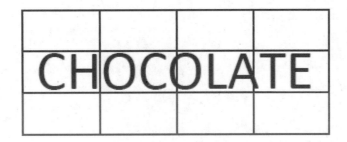

_____

_____

_____

_____

Read   **Draw**   **Write**

Name _____ Date 2-21-23

Each rectangle represents 1.

1. The shaded fractions have been decomposed into smaller units. Express the equivalent fractions in a number sentence using multiplication. The first one has been done for you.

a.

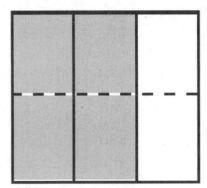

$$\frac{2}{3} = \frac{2 \times 2}{3 \times 2} = \frac{4}{6}$$

b.

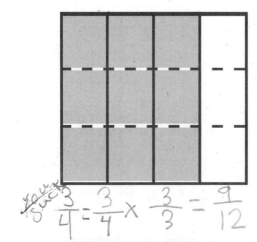

$$\frac{3}{4} = \frac{3}{4} \times \frac{3}{3} = \frac{9}{12}$$

c.

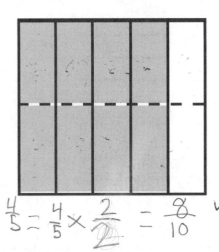

$$\frac{4}{5} = \frac{4}{5} \times \frac{2}{2} = \frac{8}{10}$$

d.

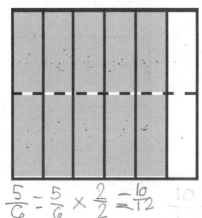

$$\frac{5}{6} = \frac{5}{6} \times \frac{2}{2} = \frac{10}{12}$$

2. Decompose the shaded fractions into smaller units, as given below. Express the equivalent fractions in a number sentence using multiplication.

a. Decompose into tenths.

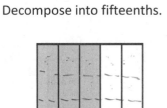

$$\frac{3}{5} = \frac{3}{5} \times \frac{2}{2} = \frac{6}{10}$$

b. Decompose into fifteenths.

$$\frac{3}{5} = \frac{3}{5} \times \frac{3}{3} = \frac{9}{15}$$

Lesson 8:   Use the area model and multiplication to show the equivalence of two fractions.

53

© 2018 Great Minds®. eureka-math.org

3. Draw area models to prove that the following number sentences are true.

a. $\frac{2}{5} = \frac{4}{10}$

    True

b. $\frac{2}{3} = \frac{8}{12}$

c. $\frac{3}{6} = \frac{6}{12}$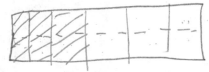

    True

d. $\frac{4}{6} = \frac{8}{12}$

4. Use multiplication to find an equivalent fraction for each fraction below.

a. $\frac{3}{4}$

b. $\frac{4}{5}$

c. $\frac{7}{6}$

d. $\frac{12}{7}$

5. Determine which of the following are true number sentences. Correct those that are false by changing the right-hand side of the number sentence.

a. $\frac{4}{3} = \frac{8}{9}$

b. $\frac{5}{4} = \frac{10}{8}$

c. $\frac{4}{5} = \frac{12}{10}$

d. $\frac{4}{6} = \frac{12}{18}$

Lesson 8: Use the area model and multiplication to show the equivalence of two fractions.

© 2018 Great Minds®. eureka-math.org

EUREKA
MATH

What fraction of a foot is 1 inch? What fraction of a foot is 3 inches? (Hint: 12 inches = 1 foot.)
Draw a tape diagram to model your work.

_____

_____

_____

_____

Read   **Draw**   **Write**

Lesson 9:   Use the area model and division to show the equivalence of
             two fractions.

Name _____   Date _____

Each rectangle represents 1.

1. Compose the shaded fractions into larger fractional units. Express the equivalent fractions in a number sentence using division. The first one has been done for you.

a.

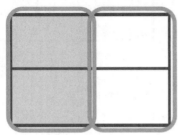

$$\frac{2}{4} = \frac{2 \div 2}{4 \div 2} = \frac{1}{2}$$

b.

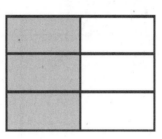

c.

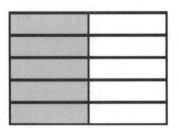

d.

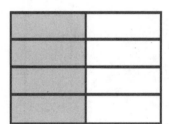

EUREKA MATH·

Lesson 9:   Use the area model and division to show the equivalence of two fractions.

59

© 2018 Great Minds®. eureka-math.org

2. Compose the shaded fractions into larger fractional units. Express the equivalent fractions in a number sentence using division.

a.
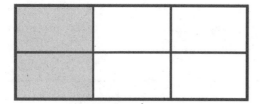

b.
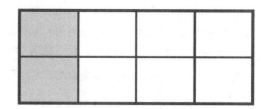

c.
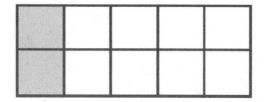

d.
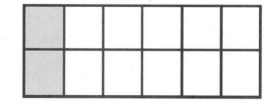

e. What happened to the size of the fractional units when you composed the fraction?

f. What happened to the total number of units in the whole when you composed the fraction?

Lesson 9:   Use the area model and division to show the equivalence of two fractions.

EUREKA
MATH

3. a. In the first area model, show 2 sixths. In the second area model, show 3 ninths. Show how both fractions can be renamed as the same unit fraction.

   b. Express the equivalent fractions in a number sentence using division.

4. a. In the first area model, show 2 eighths. In the second area model, show 3 twelfths. Show how both fractions can be composed, or renamed, as the same unit fraction.

   b. Express the equivalent fractions in a number sentence using division.

Lesson 9:   Use the area model and division to show the equivalence of two fractions.

61

© 2018 Great Minds®. eureka-math.org

Name _____   Date _____

a. In the first area model, show 2 sixths. In the second area model, show 4 twelfths. Show how both fractions can be composed, or renamed, as the same unit fraction.

b. Express the equivalent fractions in a number sentence using division.

Lesson 9:   Use the area model and division to show the equivalence of two fractions.

© 2018 Great Minds®. eureka-math.org

63

Nuri spent $\frac{9}{12}$ of his money on a book and the rest of his money on a pencil.

a. Express how much of his money he spent on the pencil in fourths.

b. Nuri started with $1. How much did he spend on the pencil?

Read　　　　**Draw**　　　　**Write**

Lesson 10:　　Use the area model and division to show the equivalence of two
fractions.

65

Name _____   Date _____

Each rectangle represents 1.

1. Compose the shaded fraction into larger fractional units. Express the equivalent fractions in a number sentence using division. The first one has been done for you.

a.

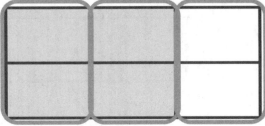

$$\frac{4}{6} = \frac{4 \div 2}{6 \div 2} = \frac{2}{3}$$

b.

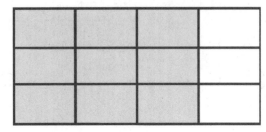

c.

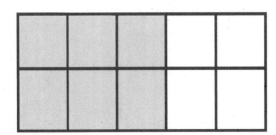

d.

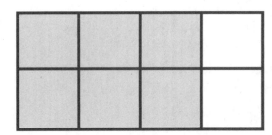

2. Compose the shaded fractions into larger fractional units. Express the equivalent fractions in a number sentence using division.

a.

b.

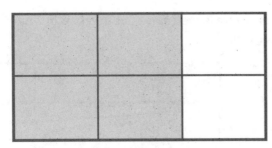

3. Draw an area model to represent each number sentence below.

a. $\dfrac{4}{10} = \dfrac{4 \div 2}{10 \div 2} = \dfrac{2}{5}$

b. $\dfrac{6}{9} = \dfrac{6 \div 3}{9 \div 3} = \dfrac{2}{3}$

Lesson 10:   Use the area model and division to show the equivalence of two fractions.

EUREKA
MATH

4. Use division to rename each fraction given below. Draw a model if that helps you. See if you can use the largest common factor.

a. $\frac{4}{8}$

b. $\frac{12}{16}$

c. $\frac{12}{20}$

d. $\frac{16}{20}$

EUREKA
MATH

Lesson 10:   Use the area model and division to show the equivalence of two
fractions.

© 2018 Great Minds®. eureka-math.org

69

Name _____   Date _____

Draw an area model to show why the fractions are equivalent. Show the equivalence in a number sentence using division.

$$\frac{4}{10} = \frac{2}{5}$$

Lesson 10:   Use the area model and division to show the equivalence of two fractions.

© 2018 Great Minds®. eureka-math.org

71

Kelly was baking bread but could only find her $\frac{1}{8}$-cup measuring cup. She needs $\frac{1}{4}$ cup sugar, $\frac{3}{4}$ cup whole wheat flour, and $\frac{1}{2}$ cup all-purpose flour. How many $\frac{1}{8}$ cups will she need for each ingredient?

_____

_____

_____

_____

Read   **Draw**   **Write**

EUREKA MATH

Lesson 11:   Explain fraction equivalence using a tape diagram and the number line, and relate that to the use of multiplication and division.

© 2018 Great Minds®. eureka-math.org

73

Name _____   Date _____

1. Label each number line with the fractions shown on the tape diagram. Circle the fraction that labels the point on the number line that also names the shaded part of the tape diagram.

a.

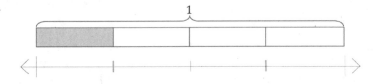

b.

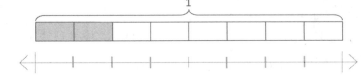

c.

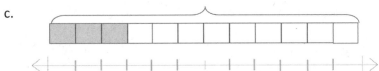

EUREKA
MATH

Lesson 11:   Explain fraction equivalence using a tape diagram and the number line, and relate that to the use of multiplication and division.

© 2018 Great Minds®. eureka-math.org

75

2. Write number sentences using multiplication to show:

    a. The fraction represented in 1(a) is equivalent to the fraction represented in 1(b).

    b. The fraction represented in 1(a) is equivalent to the fraction represented in 1(c).

3. Use each shaded tape diagram below as a ruler to draw a number line. Mark each number line with the fractional units shown on the tape diagram, and circle the fraction that labels the point on the number line that also names the shaded part of the tape diagram.

    a.

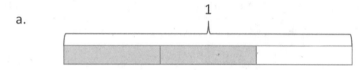

    b.

    c.

Lesson 11:   Explain fraction equivalence using a tape diagram and the number
                             line, and relate that to the use of multiplication and division.

                    © 2018 Great Minds®. eureka-math.org

EUREKA
MATH

4. Write number sentences using division to show:

    a. The fraction represented in 3(a) is equivalent to the fraction represented in 3(b).

    b. The fraction represented in 3(a) is equivalent to the fraction represented in 3(c).

5. a. Partition a number line from 0 to 1 into fifths. Decompose $\frac{2}{5}$ into 4 equal lengths.

    b. Write a number sentence using multiplication to show what fraction represented on the number line is equivalent to $\frac{2}{5}$.

    c. Write a number sentence using division to show what fraction represented on the number line is equivalent to $\frac{2}{5}$.

Lesson 11:   Explain fraction equivalence using a tape diagram and the number line, and relate that to the use of multiplication and division.

77

© 2018 Great Minds®. eureka-math.org

Name _____   Date _____

1. Partition a number line from 0 to 1 into sixths. Decompose $\frac{2}{6}$ into 4 equal lengths.

2. Write a number sentence using multiplication to show what fraction represented on the number line is equivalent to $\frac{2}{6}$.

3. Write a number sentence using division to show what fraction represented on the number line is equivalent to $\frac{2}{6}$.

Lesson 11:   Explain fraction equivalence using a tape diagram and the number line, and relate that to the use of multiplication and division.

79

Plot $\frac{1}{4}, \frac{4}{5}$, and $\frac{5}{8}$ on a number line, and compare the three points.

_____

_____

_____

_____

Read   **Draw**   **Write**

Lesson 12:   Reason using benchmarks to compare two fractions on the
number line.

Name _____ Date 2-24-23

1. a. Plot the following points on the number line without measuring.

i. $\frac{1}{3}$ ii. $\frac{5}{6}$ iii. $\frac{7}{12}$

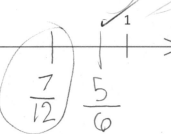

0   ✓ $\frac{1}{2}$ ✓ 1

$\frac{1}{3}$ $\frac{7}{12}$ $\frac{5}{6}$

b. Use the number line in Part (a) to compare the fractions by writing >, <, or = on the lines.

i. $\frac{7}{12}$  $\frac{1}{2}$ ✓   ii. $\frac{7}{12}$  $\frac{5}{6}$

2. a. Plot the following points on the number line without measuring.

i. $\frac{11}{12}$ ii. $\frac{1}{4}$ iii. $\frac{3}{8}$

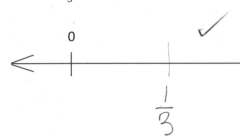

0                                $\frac{1}{2}$ ✓ 1

$\frac{1}{4}$ $\frac{1}{4}$ $\frac{3}{8}$ $\frac{3}{8}$ $\frac{11}{12}$

b. Select two fractions from Part (a), and use the given number line to compare them by writing >, <, or =.

$\frac{1}{12}$ 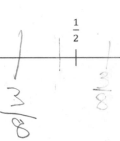 $\frac{2}{5}$ ✓

c. Explain how you plotted the points in Part (a).

$\frac{11}{12}$ is almost 1 whole. $2\frac{3}{8}$ is because 2 on the

~~$\frac{11}{12}$ is smaller then~~ $\frac{3}{8}$

top and 1 is on the top so that how

~~$\frac{}{}$ is bigger~~

3. Compare the fractions given below by writing > or < on the lines.

Give a brief explanation for each answer referring to the benchmarks 0, $\frac{1}{2}$, and 1.

a. $\frac{1}{2}$ _____ $<$ _____ $\frac{3}{4}$

b. $\frac{1}{2}$ _____ $<$ _____ $\frac{7}{8}$

c. $\frac{2}{3}$ _____ $>$ _____ $\frac{2}{5}$

d. $\frac{9}{10}$ _____ $>$ _____ $\frac{3}{5}$

e. $\frac{2}{3}$ _____ $=$ _____ $\frac{7}{8}$

f. $\frac{1}{3}$ _____ $<$ _____ $\frac{2}{4}$

g. $\frac{2}{3}$ _____ $>$ _____ $\frac{5}{10}$

h. $\frac{11}{12}$ _____ $>$ _____ $\frac{2}{5}$

i. $\frac{49}{100}$ _____ $<$ _____ $\frac{51}{100}$

j. $\frac{7}{16}$ _____ $<$ _____ $\frac{51}{100}$

Lesson 12:   Reason using benchmarks to compare two fractions on the number line.

EUREKA
MATH

Application Problem

1.

2.

_____

number line

Lesson 12: Reason using benchmarks to compare two fractions on the number line.

© 2018 Great Minds®. eureka-math.org

 87

Mr. and Mrs. Reynolds went for a run. Mr. Reynolds ran for $\frac{6}{10}$ mile. Mrs. Reynolds ran for $\frac{2}{5}$ mile. Who ran farther? Explain how you know. Use the benchmarks 0, $\frac{1}{2}$, and 1 to explain your answer.

_____

_____

_____

_____

Read   **Draw**   **Write**

Lesson 13:   Reason using benchmarks to compare two fractions on the number line.

© 2018 Great Minds®. eureka-math.org

89

Name _____   Date _____

1. Place the following fractions on the number line given.

   a. $\frac{4}{3}$　　　　　　　　　b. $\frac{11}{6}$　　　　　　　　c. $\frac{17}{12}$

   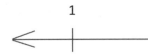

2. Use the number line in Problem 1 to compare the fractions by writing >, <, or = on the lines.

   a. $1\frac{5}{6}$ _____ $1\frac{5}{12}$　　　　　b. $1\frac{1}{3}$ _____ $1\frac{5}{12}$

3. Place the following fractions on the number line given.

   a. $\frac{11}{8}$　　　　　　　　　b. $\frac{7}{4}$　　　　　　　　c. $\frac{15}{12}$

4. Use the number line in Problem 3 to explain the reasoning you used when determining whether $\frac{11}{8}$ or $\frac{15}{12}$ is greater.

Lesson 13:　　Reason using benchmarks to compare two fractions on the number line.

91

© 2018 Great Minds®. eureka-math.org

EUREKA MATH®

5. Compare the fractions given below by writing > or < on the lines. Give a brief explanation for each answer referring to benchmarks.

a. $\dfrac{3}{8}$ _____ $\dfrac{7}{12}$

b. $\dfrac{5}{12}$ _____ $\dfrac{7}{8}$

c. $\dfrac{8}{6}$ _____ $\dfrac{11}{12}$

d. $\dfrac{5}{12}$ _____ $\dfrac{1}{3}$

e. $\dfrac{7}{5}$ _____ $\dfrac{11}{10}$

f. $\dfrac{5}{4}$ _____ $\dfrac{7}{8}$

g. $\dfrac{13}{12}$ _____ $\dfrac{9}{10}$

h. $\dfrac{6}{8}$ _____ $\dfrac{5}{4}$

i. $\dfrac{8}{12}$ _____ $\dfrac{8}{4}$

j. $\dfrac{7}{5}$ _____ $\dfrac{16}{10}$

Lesson 13:   Reason using benchmarks to compare two fractions on the number line.

EUREKA
MATH

Name _____   Date _____

1. Place the following fractions on the number line given.

   a. $\frac{5}{4}$ b. $\frac{10}{7}$ c. $\frac{16}{9}$

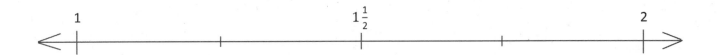

2. Compare the fractions using >, <, or =.

   a. $\frac{5}{4}$ _____ $\frac{10}{7}$ b. $\frac{5}{4}$ _____ $\frac{16}{9}$ c. $\frac{16}{9}$ _____ $\frac{10}{7}$

EUREKA MATH®

Lesson 13:   Reason using benchmarks to compare two fractions on the number line.

© 2018 Great Minds®. eureka-math.org

93

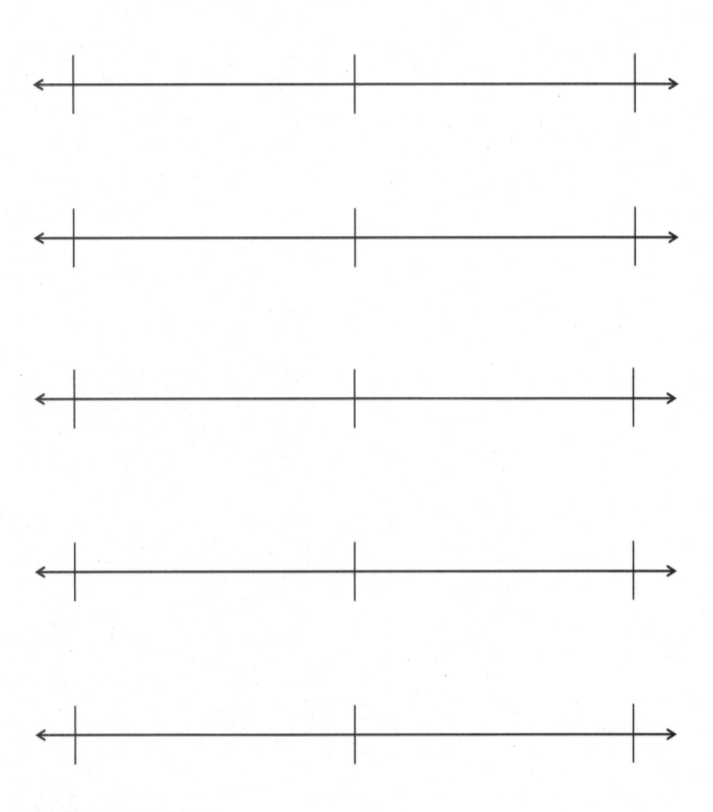

blank number lines with midpoint

Lesson 13:   Reason using benchmarks to compare two fractions on the
number line.

© 2018 Great Minds®. eureka-math.org

95

Compare $\frac{4}{5}$, $\frac{3}{4}$, and $\frac{9}{10}$ using <, >, or =. Explain your reasoning using a benchmark.

_____

_____

_____

_____

Read   **Draw**   **Write**

Name _____   Date 2-27-23 ♡♡

1. Compare the pairs of fractions by reasoning about the size of the units. Use >, <, or =.

   a. 1 fourth ___>___ 1 fifth   b. 3 fourths ___>___ 3 fifths

   c. 1 tenth ___>___ 1 twelfth   d. 7 tenths ___>___ 7 twelfths

2. Compare by reasoning about the following pairs of fractions with the same or related numerators. Use >, <, or =. Explain your thinking using words, pictures, or numbers. Problem 2(b) has been done for you.

   a. $\frac{3}{5}$ ___<___ $\frac{3}{4}$ because $\frac{1}{5} < \frac{1}{4}$

   b. $\frac{2}{5} < \frac{4}{9}$

   because $\frac{2}{5} = \frac{4}{10}$
   4 tenths is less
   than 4 ninths because
   tenths are smaller than
   ninths.

   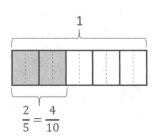

   1

   $\frac{2}{5} = \frac{4}{10}$

   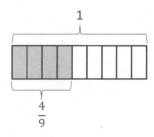

   1

   $\frac{4}{9}$

   c. $\frac{7}{11}$ ___>___ $\frac{7}{13}$

   d. $\frac{6}{7}$ ___>___ $\frac{12}{15}$

   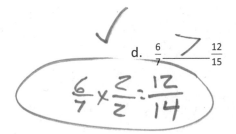

   $\frac{6}{7} \times \frac{2}{2} = \frac{12}{14}$

   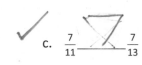

   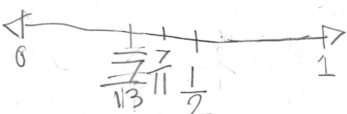

   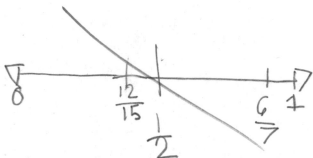

EUREKA
MATH

© 2018 Great Minds®. eureka-math.org

3. Draw two tape diagrams to model each pair of the following fractions with related denominators. Use >, <, or = to compare.

a. 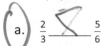 $\frac{2}{3}$ > $\frac{5}{6}$

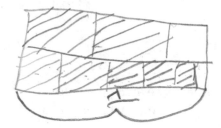

b. $\frac{3}{4}$ < $\frac{7}{8}$

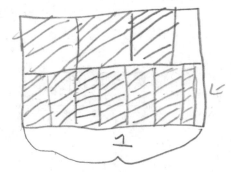

 c. $1\frac{3}{4}$ > $1\frac{7}{12}$

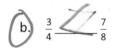

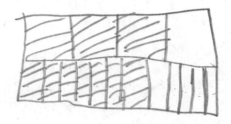

EUREKA MATH

4. Draw one number line to model each pair of fractions with related denominators. Use >, <, or = to compare.

a. $\frac{2}{3}$ _____ $\frac{5}{6}$

b. $\frac{3}{8}$ _____ $\frac{1}{4}$

c. $\frac{2}{6}$ _____ $\frac{5}{12}$

d. $\frac{8}{9}$ _____ $\frac{2}{3}$

5. Compare each pair of fractions using >, <, or =. Draw a model if you choose to.

a. $\frac{3}{4}$ _____ $\frac{3}{7}$

b. $\frac{4}{5}$ _____ $\frac{8}{12}$

c. $\frac{7}{10}$ _____ $\frac{3}{5}$

d. $\frac{2}{3}$ _____ $\frac{11}{15}$

e. $\frac{3}{4}$ _____ $\frac{11}{12}$

f. $\frac{7}{3}$ _____ $\frac{7}{4}$

g. $1\frac{1}{3}$ _____ $1\frac{2}{9}$

h. $1\frac{2}{3}$ _____ $1\frac{4}{7}$

EUREKA
MATH®

6. Timmy drew the picture to the right and claimed that $\frac{2}{3}$ is less than $\frac{7}{12}$. Evan says he thinks $\frac{2}{3}$ is greater than $\frac{7}{12}$. Who is correct? Support your answer with a picture.

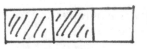

$\frac{2}{3}$

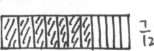

$\frac{7}{12}$

EUREKA
MATH

Jamal ran $\frac{2}{3}$ mile. Ming ran $\frac{2}{4}$ mile. Laina ran $\frac{7}{12}$ mile. Who ran the farthest? What do you think is the easiest way to determine the answer to this question?

_____

_____

_____

_____

Read   **Draw**   **Write**

Lesson 15:   Find common units or number of units to compare two fractions.

105

Name _____ Date 2-28-23

1. Draw an area model for each pair of fractions, and use it to compare the two fractions by writing >, <, or = on the line. The first two have been partially done for you. Each rectangle represents 1.

a. $\frac{1}{2}$ ____ < ____ $\frac{2}{3}$

$\frac{1 \times 3}{2 \times 3} = \frac{3}{6}$

$\frac{2 \times 2}{3 \times 2} = \frac{4}{6}$

b. $\frac{4}{5}$ > $\frac{3}{4}$

$\frac{4}{5} \times \frac{4}{4} = \frac{16}{20}$

$\frac{3}{4} \times \frac{5}{5} = \frac{15}{20}$

c. $\frac{3}{5}$ > $\frac{4}{7}$

$\frac{3}{5} \times \frac{7}{7} = \frac{21}{35}$

$\frac{4}{7} \times \frac{5}{5} = \frac{20}{35}$

d. $\frac{3}{7}$ > $\frac{2}{6}$

$\frac{3}{7} \times \frac{6}{6} = \frac{18}{42}$

$\frac{2}{6} \times \frac{7}{7} = \frac{14}{42}$

e. $\frac{5}{8}$ < $\frac{6}{9}$

$\frac{5}{8} \times \frac{9}{9} = \frac{45}{72}$

$\frac{6}{9} \times \frac{8}{8} = \frac{48}{72}$

f. $\frac{2}{3}$ ____ $\frac{3}{4}$

$\frac{2}{3} \times \frac{4}{4} = \frac{18}{ }$

$\frac{3}{4} \times \frac{3}{3} = \frac{16}{16}$

Lesson 15:   Find common units or number of units to compare two fractions.

107

© 2018 Great Minds®. eureka-math.org

2. Rename the fractions, as needed, using multiplication in order to compare each pair of fractions by writing >, <, or =.

a. $\dfrac{3}{5}$ _____ $\dfrac{5}{6}$

b. $\dfrac{2}{6}$ _____ $\dfrac{3}{8}$

c. $\dfrac{7}{5}$ _____ $\dfrac{10}{8}$

d. $\dfrac{4}{3}$ _____ $\dfrac{6}{5}$

3. Use any method to compare the fractions. Record your answer using >, <, or =.

a. $\dfrac{3}{4}$ _____ $\dfrac{7}{8}$

b. $\dfrac{6}{8}$ _____ $\dfrac{3}{5}$

c. $\dfrac{6}{4}$ _____ $\dfrac{8}{6}$

d. $\dfrac{8}{5}$ _____ $\dfrac{9}{6}$

EUREKA
MATH

4. Explain two ways you have learned to compare fractions. Provide evidence using words, pictures, or numbers.

Addison

- I have no clue

Same after the brack i lost my mind

3 6
4 5
5 4
6 3
7 2
8 1
9 0

EUREKA MATH

Lesson 15:   Find common units or number of units to compare two fractions.

© 2018 Great Minds®. eureka-math.org

109

Keisha ran $\frac{5}{6}$ mile in the morning and $\frac{2}{3}$ mile in the afternoon. Did Keisha run farther in the morning or in the afternoon? Explain.

_____

_____

_____

_____

Read   **Draw**   **Write**

Lesson 16:   Use visual models to add and subtract two fractions with the same
units.

113

Name _____   Date 3-1-23

1. Solve.

    a. 3 fifths − 1 fifth = __2__ Fifths

    b. 5 fifths − 3 fifths = __2__ Fifths

    c. 3 halves − 2 halves = __1 halve__

    d. 6 fourths − 3 fourths = __3 Fourths__

2. Solve.

    a. $\frac{5}{6} - \frac{3}{6} = \frac{2}{6}$

    b. $\frac{6}{8} - \frac{4}{8} = \frac{2}{8}$

    c. $\frac{3}{10} - \frac{3}{10} = \frac{0}{10}$

    d. $\frac{5}{5} - \frac{4}{5} = \frac{1}{5}$

    e. $\frac{5}{4} - \frac{4}{4} = \frac{1}{4}$

    f. $\frac{5}{4} - \frac{3}{4} = \frac{2}{4}$

3. Solve. Use a number bond to show how to convert the difference to a mixed number. Problem (a) has been completed for you.

    a. $\frac{12}{8} - \frac{3}{8} = \frac{9}{8} = 1\frac{1}{8}$

    b. $\frac{12}{6} - \frac{5}{6} = \frac{7}{6} = 1\frac{1}{6}$

    c. $\frac{9}{5} - \frac{3}{5} = \frac{6}{5} = 1\frac{1}{6}$

    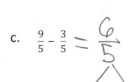

    d. $\frac{14}{8} - \frac{3}{8} = \frac{11}{8} = 1\frac{3}{8}$

    e. $\frac{8}{4} - \frac{2}{4} = \frac{6}{4} = 1\frac{R}{6}$

    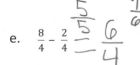
    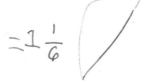

    f. $\frac{15}{10} - \frac{3}{10} = \frac{12}{10} = 1\frac{2}{10}$

    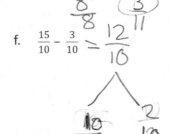

4.   Solve. Write the sum in unit form.

   a.   2 fourths + 1 fourth = _____   b.   4 fifths + 3 fifths = _____

5.   Solve.

   a. $\frac{2}{8} + \frac{5}{8}$ b. $\frac{4}{12} + \frac{5}{12}$

6.   Solve. Use a number bond to decompose the sum. Record your final answer as a mixed number.
     Problem (a) has been completed for you.

   a. $\frac{3}{5} + \frac{4}{5} = \frac{7}{5} = 1\frac{2}{5}$ b. $\frac{4}{4} + \frac{3}{4}$

           $\frac{5}{5}$ $\frac{2}{5}$

   c. $\frac{6}{9} + \frac{6}{9}$ d. $\frac{7}{10} + \frac{6}{10}$

   e. $\frac{5}{6} + \frac{7}{6}$ f. $\frac{9}{8} + \frac{5}{8}$

7.   Solve. Use a number line to model your answer.

   a. $\frac{7}{4} - \frac{5}{4}$

   b. $\frac{5}{4} + \frac{2}{4}$

Lesson 16:   Use visual models to add and subtract two fractions with the same
                                     units.

EUREKA
MATH

Name _____ Date _____

1. Solve. Use a number bond to decompose the difference. Record your final answer as a mixed number.

    $\frac{16}{9} - \frac{5}{9}$

2. Solve. Use a number bond to decompose the sum. Record your final answer as a mixed number.

    $\frac{5}{12} + \frac{10}{12}$

Lesson 16:   Use visual models to add and subtract two fractions with the same units.

© 2018 Great Minds®. eureka-math.org

117

Name _____ Date _____

blank number lines

Lesson 16: Use visual models to add and subtract two fractions with the same units.

© 2018 Great Minds®. eureka-math.org

119

Use a number bond to show the relationship between $\frac{2}{3}$, $\frac{3}{6}$, and $\frac{5}{6}$. Then, use the fractions to write two addition and two subtraction sentences.

_____

_____

_____

_____

Read   **Draw**   **Write**

Lesson 17:   Use visual models to add and subtract two fractions with the same
               units, including subtracting from one whole.   121

© 2018 Great Minds®. eureka-math.org

Name _____   Date 3-1-23

1. Use the following three fractions to write two subtraction and two addition number sentences.

a. $\frac{8}{5}$, $\frac{2}{5}$, $\frac{10}{5}$

$$\frac{8}{5} + \frac{2}{5} = \frac{10}{5}$$

$$\frac{10}{5} - \frac{2}{5} = \frac{8}{5}$$

$$\frac{2}{5} + \frac{8}{5} = \frac{10}{5}$$

$$\frac{10}{5} - \frac{8}{5} = \frac{2}{5}$$

b. $\frac{15}{8}$, $\frac{7}{8}$, $\frac{8}{8}$

$$\frac{8}{8} + \frac{7}{8} = \frac{15}{8}$$

$$\frac{15}{8} - \frac{7}{8} = \frac{8}{8}$$

$$\frac{7}{8} + \frac{8}{8} = \frac{15}{8}$$

$$\frac{15}{8} - \frac{8}{8} = \frac{7}{8}$$

2. Solve. Model each subtraction problem with a number line, and solve by both counting up and subtracting. Part (a) has been completed for you.

a. $1 - \frac{3}{4}$

$$\frac{4}{4} - \frac{3}{4} = \frac{1}{4}$$

b. $1 - \frac{8}{10}$ $\frac{10}{10} - \frac{8}{10} = \frac{1}{10}$

c. $1 - \frac{3}{5}$

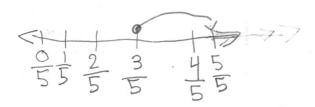

d. $1 - \frac{5}{8}$

e. $1\frac{2}{10} - \frac{7}{10}$

f. $1\frac{1}{5} - \frac{3}{5}$

Lesson 17:   Use visual models to add and subtract two fractions with the same
             units, including subtracting from one whole.

123

© 2018 Great Minds®. eureka-math.org

3. Find the difference in two ways. Use number bonds to decompose the total. Part (a) has been completed for you.

a. $1\frac{2}{5} - \frac{4}{5}$

$$\frac{5}{5} \quad \frac{2}{5}$$

$$\frac{5}{5} + \frac{2}{5} = \frac{7}{5}$$

$$\frac{7}{5} - \frac{4}{5} = \boxed{\frac{3}{5}}$$

$$\frac{5}{5} - \frac{4}{5} = \frac{1}{5}$$

$$\frac{1}{5} + \frac{2}{5} = \boxed{\frac{3}{5}}$$

b. $1\frac{3}{6} - \frac{4}{6}$

c. $1\frac{6}{8} - \frac{7}{8}$

d. $1\frac{1}{10} - \frac{7}{10}$

e. $1\frac{3}{12} - \frac{6}{12}$

Use visual models to add and subtract two fractions with the same units, including subtracting from one whole.

EUREKA MATH

Name _____   Date _____

1. Solve. Model the problem with a number line, and solve by both counting up and subtracting.

    $1 - \frac{2}{5}$

2. Find the difference in two ways. Use a number bond to show the decomposition.

    $1\frac{2}{7} - \frac{5}{7}$

Lesson 17:   Use visual models to add and subtract two fractions with the same
               units, including subtracting from one whole.

© 2018 Great Minds®. eureka-math.org

125

Name _____   Date _____

Problem A:	$\frac{1}{8} + \frac{3}{8} + \frac{4}{8}$	

Problem B:	$\frac{1}{6} + \frac{4}{6} + \frac{2}{6}$	

Problem C:	$\frac{11}{10} - \frac{4}{10} - \frac{1}{10}$	

adding and subtracting fractions

Problem D: $1 - \dfrac{3}{12} - \dfrac{5}{12}$

Problem E: $\dfrac{5}{8} + \dfrac{4}{8} + \dfrac{1}{8}$

Problem F: $1\dfrac{1}{5} - \dfrac{2}{5} - \dfrac{3}{5}$

adding and subtracting fractions

EUREKA
MATH®

Name _____   Date _3-2-23_

1. Show one way to solve each problem. Express sums and differences as a mixed number when possible.
    Use number bonds when it helps you. Part (a) is partially completed.

a. $\frac{2}{5} + \frac{3}{5} + \frac{1}{5}$ $= \frac{5}{5} + \frac{1}{5} = 1 + \frac{1}{5}$ $= $ _____	b. $\frac{3}{6} + \frac{1}{6} + \frac{3}{6} = \frac{7}{6} = 1\frac{1}{6}$	c. $\frac{5}{7} + \frac{7}{7} + \frac{2}{7} = \frac{14}{7} = 2$ 
d. $\frac{7}{8} - \frac{3}{8} - \frac{1}{8} = \frac{3}{8}$	e. $\frac{7}{9} + \frac{1}{9} + \frac{4}{9} = \frac{12}{9} = 1$ 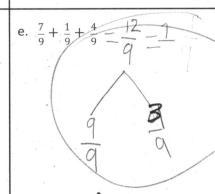	f. $\frac{4}{10} + \frac{11}{10} + \frac{5}{10} = \frac{20}{10} = 2$ 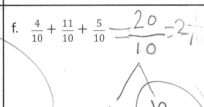
g. $1 - \frac{3}{12} - \frac{4}{12} = \frac{5}{12}$ 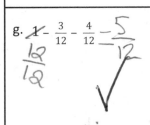	h. $1\frac{2}{3} - \frac{1}{3} - \frac{1}{3}$	i. $\frac{10}{12} + \frac{5}{12} + \frac{2}{12} + \frac{7}{12} = \frac{24}{12} = 2$ 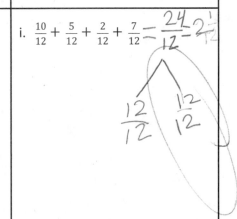

2.   Monica and Stuart used different strategies to solve $\frac{5}{8} + \frac{2}{8} + \frac{5}{8}$.

Monica's Way

$$\frac{5}{8} + \frac{2}{8} + \frac{5}{8} = \frac{7}{8} + \frac{5}{8} = \frac{8}{8} + \frac{4}{8} = 1\frac{4}{8}$$

$$\frac{1}{8} \qquad \frac{4}{8}$$

Stuart's Way

$$\frac{5}{8} + \frac{2}{8} + \frac{5}{8} = \frac{12}{8} = 1 + \frac{4}{8} = 1\frac{4}{8}$$

$$\frac{8}{8} \qquad \frac{4}{8}$$

Whose strategy do you like best? Why?

I use stuar'ts way because it is esay for me

3.   You gave one solution for each part of Problem 1. Now, for each problem indicated below, give a different solution method.

1(c) $\frac{5}{7} + \frac{7}{7} + \frac{2}{7} = \frac{14}{7} = 2$

$$\frac{7}{7} \qquad \frac{7}{7}$$

1(f) $\frac{4}{10} + \frac{11}{10} + \frac{5}{10} = \frac{20}{10} = 2$

$$\frac{10}{10} \qquad \frac{10}{10}$$

1(g) $1 - \frac{3}{12} - \frac{4}{12}$

$$\frac{12}{12} - \frac{3}{12} = 9 \quad \frac{4}{12} = 5$$

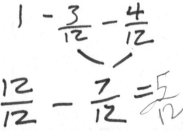

$1 - \frac{3}{12} - \frac{4}{12}$

$$\frac{12}{12} - \frac{7}{12} = \frac{5}{12}$$

EUREKA MATH

Fractions are all around us! Make a list of times that you have used fractions, heard fractions, or seen fractions. Be ready to share your ideas.

_____

_____

_____

_____

Read   Draw   Write

Name _____   Date 3-3-23

Use the RDW process to solve.

1. Sue ran $\frac{9}{10}$ mile on Monday and $\frac{7}{10}$ mile on Tuesday. How many miles did Sue run in the 2 days?

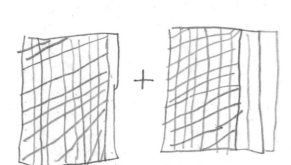 +

$$\frac{9}{10} + \frac{7}{10} = \frac{16}{10}$$

2. Mr. Salazar cut his son's birthday cake into 8 equal pieces. Mr. Salazar, Mrs. Salazar, and the birthday boy each ate 1 piece of cake. What fraction of the cake was left?

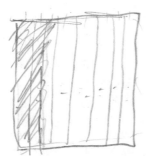

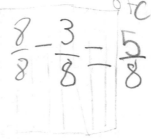

$\frac{5}{8}$ Left of the cake was

$$\frac{8}{8} - \frac{3}{8} = \frac{5}{8}$$

3. Maria spent $\frac{4}{7}$ of her money on a book and saved the rest. What fraction of her money did Maria save?

4. Mrs. Jones had $1\frac{4}{8}$ pizzas left after a party. After giving some to Gary, she had $\frac{7}{8}$ pizza left. What fraction of a pizza did she give Gary?

5. A baker had 2 pans of corn bread. He served $1\frac{1}{4}$ pans. What fraction of a pan was left?

6. Marius combined $\frac{4}{8}$ gallon of lemonade, $\frac{3}{8}$ gallon of cranberry juice, and $\frac{6}{8}$ gallon of soda water to make punch for a party. How many gallons of punch did he make in all?

Lesson 19:   Solve word problems involving addition and subtraction of fractions.

EUREKA
MATH

Krista drank $\frac{3}{16}$ of the water in her water bottle in the morning, $\frac{5}{16}$ in the afternoon, and $\frac{3}{16}$ in the evening. What fraction of water was left at the end of the day?

_____

_____

_____

_____

Read   **Draw**   **Write**

Lesson 20:   Use visual models to add two fractions with related units using the
denominators 2, 3, 4, 5, 6, 8, 10, and 12.

© 2018 Great Minds®. eureka-math.org

Name _____   Date _____

1. Use a tape diagram to represent each addend. Decompose one of the tape diagrams to make like units. Then, write the complete number sentence. Part (a) is partially completed.

   a. $\frac{1}{4} + \frac{1}{8}$

   b. $\frac{1}{4} + \frac{1}{12}$

   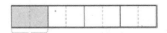   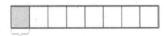

   $$\frac{}{8} + \frac{}{8} = \frac{}{8}$$

   c. $\frac{2}{6} + \frac{1}{3}$

   d. $\frac{1}{2} + \frac{3}{8}$

   e. $\frac{3}{10} + \frac{3}{5}$

   f. $\frac{2}{3} + \frac{2}{9}$

EUREKA
MATH

Lesson 20: Use visual models to add two fractions with related units using the denominators 2, 3, 4, 5, 6, 8, 10, and 12.

141

© 2018 Great Minds®. eureka-math.org

2. Estimate to determine if the sum is between 0 and 1 or 1 and 2. Draw a number line to model the addition. Then, write a complete number sentence. Part (a) has been completed for you.

a. $\frac{1}{2} + \frac{1}{4}$ $\frac{2}{4} + \frac{1}{4} = \frac{3}{4}$

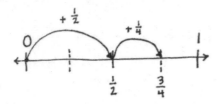

b. $\frac{1}{2} + \frac{4}{10}$

c. $\frac{6}{10} + \frac{1}{2}$

d. $\frac{2}{3} + \frac{3}{6}$

e. $\frac{3}{4} + \frac{6}{8}$

f. $\frac{4}{10} + \frac{6}{5}$

3. Solve the following addition problem without drawing a model. Show your work.

$$\frac{2}{3} + \frac{4}{6}$$

Lesson 20: Use visual models to add two fractions with related units using the denominators 2, 3, 4, 5, 6, 8, 10, and 12.

© 2018 Great Minds®. eureka-math.org

EUREKA
MATH®

Name _____   Date _____

1. Draw a number line to model the addition. Solve, and then write a complete number sentence.

$$\frac{5}{8} + \frac{2}{4}$$

2. Solve without drawing a model.

$$\frac{3}{4} + \frac{1}{2}$$

Lesson 20:   Use visual models to add two fractions with related units using the
denominators 2, 3, 4, 5, 6, 8, 10, and 12.

143

© 2018 Great Minds®. eureka-math.org

Two-fifths liter of chemical A was added to $\frac{7}{10}$ liter of chemical B to make chemical C. How many liters of chemical C are there?

_____

_____

_____

_____

Read   **Draw**   **Write**

Lesson 21:   Use visual models to add two fractions with related units using the
                denominators 2, 3, 4, 5, 6, 8, 10, and 12.

145

© 2018 Great Minds®. eureka-math.org

Name _____ Date 3-6-23

1. Draw a tape diagram to represent each addend. Decompose one of the tape diagrams to make like units. Then, write a complete number sentence. Use a number bond to write each sum as a mixed number.

   a. $\frac{3}{4} + \frac{1}{2}$

   b. $\frac{2}{3} + \frac{3}{6}$

   c. $\frac{5}{6} + \frac{1}{3} = \frac{5}{6} + \frac{2}{6} = \frac{7}{6}$

   $\frac{1}{3} \times \frac{2}{2} = \frac{2}{6}$

   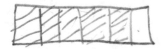

   d. $\frac{4}{5} + \frac{7}{10} = \frac{7}{10} + \frac{8}{10} = \frac{15}{10} = 1\frac{5}{10}$

   $\frac{4}{5} \times \frac{2}{2} = \frac{8}{10}$

   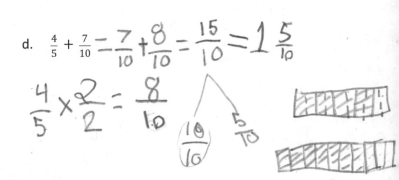

2. Draw a number line to model the addition. Then, write a complete number sentence. Use a number bond to write each sum as a mixed number.

   a. $\frac{1}{2} + \frac{3}{4}$

   b. $\frac{1}{2} + \frac{6}{8} = \frac{4}{8} + \frac{6}{8} = \frac{10}{8}$

   $\frac{1}{2} \times \frac{4}{4} = \frac{4}{8}$

   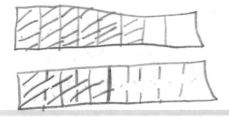

EUREKA
MATH®

Lesson 21: Use visual models to add two fractions with related units using the
denominators 2, 3, 4, 5, 6, 8, 10, and 12.

147

© 2018 Great Minds®. eureka-math.org

c. $\frac{7}{10} + \frac{3}{5}$ $= \frac{6}{10} + \frac{7}{10} = \frac{13}{10} = 1\frac{3}{10}$

$\frac{3}{5} \times \frac{2}{2} = \frac{6}{10}$ $\overset{10}{\underset{\frac{10}{10} \quad \frac{3}{10}}{\wedge}}$

d. $\frac{2}{3} + \frac{5}{6}$ $= \frac{4}{6} \times \frac{5}{6} = \frac{9}{6} = 1\frac{3}{6}$

$\frac{2}{3} \times \frac{2}{2} = \frac{4}{6}$ $\overset{}{\underset{\frac{6}{6} \qquad \frac{3}{6}}{\wedge}}$

3. Solve. Write the sum as a mixed number. Draw a model if needed.

a. $\frac{3}{4} + \frac{2}{8}$ $= \frac{6}{8} + \frac{2}{8} = \frac{8}{8} = 1$

$\frac{3}{4} \times \frac{2}{2} = \frac{6}{8}$

b. $\frac{4}{6} + \frac{1}{2}$ $= \frac{3}{6} + \frac{4}{8} = \frac{7}{6} = 1\frac{1}{6}$

$\frac{1}{2} \times \frac{3}{3} = \frac{3}{6}$ $\overset{}{\underset{\frac{6}{6} \qquad \frac{1}{6}}{\wedge}}$

c. $\frac{4}{6} + \frac{2}{3}$ $= \frac{4}{8} + \frac{4}{6} = \frac{8}{6} = 1\frac{1}{6}$

$\frac{2}{3} \times \frac{2}{2} = \frac{4}{6}$ $\overset{}{\underset{\frac{6}{6}}{\wedge}}$

d. $\frac{8}{10} + \frac{3}{5}$ $= \frac{6}{10} + \frac{8}{10} = \frac{14}{10} = 1\frac{4}{10}$

$\frac{3}{5} \times \frac{2}{2} = \frac{6}{10}$ $\overset{}{\underset{\frac{10}{10} \qquad \frac{4}{10}}{\wedge}}$

e. $\frac{5}{8} + \frac{3}{4}$ $= \frac{6}{8} + \frac{5}{8} = \frac{11}{8} = 1\frac{1}{8}$

$\frac{3}{4} \times \frac{2}{2} = \frac{6}{8}$ $\overset{}{\underset{\frac{8}{8} \qquad \frac{1}{8}}{\wedge}}$

f. $\frac{5}{8} + \frac{2}{4}$ $=$

$\frac{2}{4} \times \frac{2}{2} = \frac{4}{8}$

g. $\frac{1}{2} + \frac{5}{8}$

h. $\frac{3}{10} + \frac{4}{5}$

Lesson 21:   Use visual models to add two fractions with related units using the denominators 2, 3, 4, 5, 6, 8, 10, and 12.

© 2018 Great Minds®. eureka-math.org

EUREKA
MATH

Winnie went shopping and spent $\frac{2}{5}$ of the money that was on a gift card. What fraction of the money was left on the card? Draw a number line and a number bond to help show your thinking.

_____

_____

_____

_____

Read   **Draw**   **Write**

Lesson 22: Add a fraction less than 1 to, or subtract a fraction less than 1 from, a whole number using decomposition and visual models.

151

Name _____   Date 3-9-23

1. Draw a tape diagram to match each number sentence. Then, complete the number sentence.

a. $3 + \frac{1}{3} =$ _____

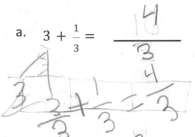

b. $4 + \frac{3}{4} =$ _____

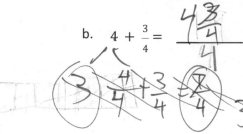

c. $3 - \frac{1}{4} =$ _____

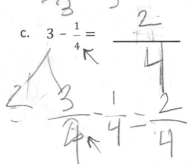

d. $5 - \frac{2}{5} = \frac{3}{5}$

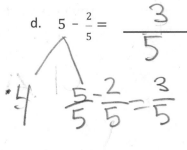

$\frac{5}{5} - \frac{2}{5} = \frac{3}{5}$

2. Use the following three numbers to write two subtraction and two addition number sentences.

a. $6, 6\frac{3}{8}, \frac{3}{8}$

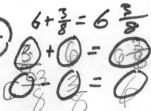

$6 + \frac{3}{8} = 6\frac{3}{8}$

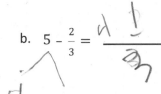

b. $\frac{4}{7}, 9, 8\frac{3}{7}$

3. Solve using a number bond. Draw a number line to represent each number sentence. The first one has been done for you.

a. $4 - \frac{1}{3} = 3\frac{2}{3}$

$4 - \frac{1}{3} = 3\frac{2}{3}$

$3 \quad \frac{3}{3}$

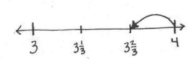

b. $5 - \frac{2}{3} =$ _____

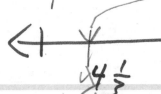

$\frac{5}{3} - \frac{2}{3} = \frac{3}{3}$

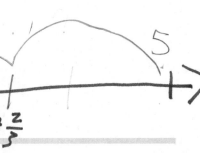

EUREKA MATH

Lesson 22: Add a fraction less than 1 to, or subtract a fraction less than 1 from, a whole number using decomposition and visual models.

153

© 2018 Great Minds®. eureka-math.org

c. $7 - \dfrac{3}{8} =$ _____   d. $10 - \dfrac{4}{10} =$ _____

4. Complete the subtraction sentences using number bonds.

a. $3 - \dfrac{1}{10} = 2\dfrac{9}{10}$

$$\dfrac{10}{10} - \dfrac{1}{10} = \dfrac{9}{10}$$

b. $5 - \dfrac{3}{4} =$ _____

$9 \quad \dfrac{4}{4} -$

c. $6 - \dfrac{5}{8} = 5\dfrac{8}{8}$

$5 \quad \dfrac{8}{8}$

$$5\dfrac{13}{8} - 5\dfrac{5}{8} = 5\dfrac{8}{8}$$

d. $7 - \dfrac{3}{9} =$ _____

e. $8 - \dfrac{6}{10} =$ _____   f. $29 - \dfrac{9}{12} =$ _____

Lesson 22:   Add a fraction less than 1 to, or subtract a fraction less than 1 from, a whole number using decomposition and visual models.

EUREKA MATH®

Mrs. Wilcox cut quilt squares and then divided them evenly into 8 piles. She decided to sew together 1 pile each night. After 5 nights, what fraction of the quilt squares was sewn together? Draw a tape diagram or a number line to model your thinking, and then write a number sentence to express your answer.

_____

_____

_____

_____

Read   **Draw**   **Write**

Lesson 23:   Add and multiply unit fractions to build fractions greater than 1 using visual models.

© 2018 Great Minds®. eureka-math.org

157

EUREKA
MATH®

Name _____   Date 3-10-23

1. Circle any fractions that are equivalent to a whole number. Record the whole number below the fraction.

   a. Count by 1 thirds. Start at 0 thirds. End at 6 thirds.

   $\frac{0}{3}$, $\frac{1}{3}$, $\frac{2}{3}$ 1

   0

   b. Count by 1 halves. Start at 0 halves. End at 8 halves.

2. Use parentheses to show how to make ones in the following number sentence.

   $\frac{1}{4} + \frac{1}{4} + \frac{1}{4} + \frac{1}{4} + \frac{1}{4} + \frac{1}{4} + \frac{1}{4} + \frac{1}{4} + \frac{1}{4} + \frac{1}{4} + \frac{1}{4} + \frac{1}{4} = 3$

3. Multiply, as shown below. Draw a number line to support your answer.

   a. $6 \times \frac{1}{3}$

   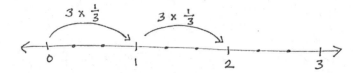

   $6 \times \frac{1}{3} = 2 \times \frac{3}{3} = 2$

   b. $6 \times \frac{1}{2}$

   c. $12 \times \frac{1}{4}$

EUREKA MATH   Lesson 23:   Add and multiply unit fractions to build fractions greater than 1 using   159
                                   visual models.

                  © 2018 Great Minds®. eureka-math.org

4. Multiply, as shown below. Write the product as a mixed number. Draw a number line to support your answer.

   a. 7 copies of 1 third

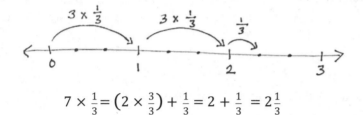

$$7 \times \frac{1}{3} = \left(2 \times \frac{3}{3}\right) + \frac{1}{3} = 2 + \frac{1}{3} = 2\frac{1}{3}$$

   b. 7 copies of 1 half

   c. $10 \times \frac{1}{4}$

   d. $14 \times \frac{1}{3}$

Lesson 23:   Add and multiply unit fractions to build fractions greater than 1 using visual models.

EUREKA
MATH®

Shelly read her book for $\frac{1}{2}$ hour each afternoon for 9 days. How many hours did Shelly spend reading in all 9 days?

_____

_____

_____

_____

Read   **Draw**   **Write**

Lesson 24:   Decompose and compose fractions greater than 1 to express them in
              various forms.

163

© 2018 Great Minds®. eureka-math.org

Name _____   Date _____

1. Rename each fraction as a mixed number by decomposing it into two parts as shown below. Model the decomposition with a number line and a number bond.

   a. $\frac{11}{3}$

   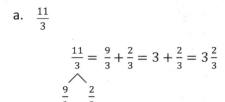

   $$\frac{11}{3} = \frac{9}{3} + \frac{2}{3} = 3 + \frac{2}{3} = 3\frac{2}{3}$$

   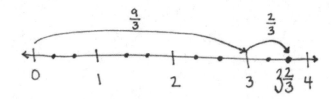

   b. $\frac{12}{5}$

   c. $\frac{13}{2}$

   d. $\frac{15}{4}$

EUREKA MATH

Lesson 24:   Decompose and compose fractions greater than 1 to express them in various forms.

165

© 2018 Great Minds®. eureka-math.org

2. Convert each fraction to a mixed number. Show your work as in the example. Model with a number line.

a. $\frac{11}{3}$

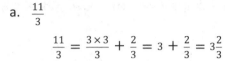

$$\frac{11}{3} = \frac{3 \times 3}{3} + \frac{2}{3} = 3 + \frac{2}{3} = 3\frac{2}{3}$$

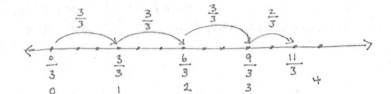

b. $\frac{9}{2}$

c. $\frac{17}{4}$

3. Convert each fraction to a mixed number.

a. $\frac{9}{4} =$	b. $\frac{17}{5} =$	c. $\frac{25}{6} =$
d. $\frac{30}{7} =$	e. $\frac{38}{8} =$	f. $\frac{48}{9} =$
g. $\frac{63}{10} =$	h. $\frac{84}{10} =$	i. $\frac{37}{12} =$

Lesson 24:   Decompose and compose fractions greater than 1 to express them in various forms.

EUREKA MATH

Mrs. Fowler knew that the perimeter of the soccer field was $\frac{1}{6}$ mile. Her goal was to walk two miles while watching her daughter's game. If she walked around the field 13 times, did she meet her goal? Explain your thinking.

_____

_____

_____

_____

Read   **Draw**   **Write**

Lesson 25:   Decompose and compose fractions greater than 1 to express them in various forms.

169

© 2018 Great Minds®. eureka-math.org

Name _____   Date 3-13-23

1. Convert each mixed number to a fraction greater than 1. Draw a number line to model your work.

a. $3\frac{1}{4}$

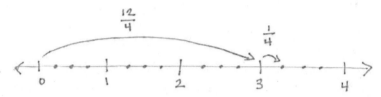

$$3\frac{1}{4} = 3 + \frac{1}{4} = \frac{12}{4} + \frac{1}{4} = \frac{13}{4}$$

b. $2\frac{4}{5}$ =

$$2\frac{4}{5} = \left(2 \times \frac{5}{5}\right) + \frac{4}{5} = \frac{10}{5} + \frac{4}{5} = \frac{14}{5}$$

I hate math

c. $3\frac{5}{8}$ =

$$3\frac{5}{8} = \left(3 \times \frac{8}{8}\right) + \frac{5}{8} = \frac{29}{8} \checkmark = \frac{5}{6} = \frac{34}{8}$$

d. $4\frac{4}{10}$ =

$$4\frac{4}{10} = \left(4 \times \frac{10}{10}\right) + \frac{4}{10} = \frac{44}{10} \checkmark$$

e. $4\frac{7}{9}$ =

$$4\frac{7}{9} = \left(4 \times \frac{9}{9}\right) + \frac{7}{9} = \frac{43}{9}$$

Lesson 25:   Decompose and compose fractions greater than 1 to express them in various forms.

171

© 2018 Great Minds®. eureka-math.org

2. Convert each mixed number to a fraction greater than 1. Show your work as in the example.
   (Note: $3 \times \frac{4}{4} = \frac{3 \times 4}{4}$.)

   a. $3\frac{3}{4}$

   $$3\frac{3}{4} = 3 + \frac{3}{4} = \left(3 \times \frac{4}{4}\right) + \frac{3}{4} = \frac{12}{4} + \frac{3}{4} = \frac{15}{4}$$

   b. $4\frac{1}{3} = \left(4 \times \frac{3}{3}\right) + \frac{1}{3} = \frac{13}{3}$

   c. $4\frac{3}{5} = \left(4 \times \frac{5}{5}\right) + \frac{3}{5} = \frac{23}{5}$

   d. $4\frac{6}{8}$ $\left(4 \times \frac{8}{8}\right) + \frac{6}{8} = \frac{38}{8}$

3. Convert each mixed number to a fraction greater than 1.

a. $2\frac{3}{4} = \left(2 \times \frac{4}{4}\right) + \frac{3}{4} = \frac{11}{4}$	b. $2\frac{2}{5} =$	c. $3\frac{3}{6} =$
d. $3\frac{3}{8} =$	e. $3\frac{1}{10} =$	f. $4\frac{3}{8} =$
g. $5\frac{2}{3} =$	h. $6\frac{1}{2} =$	i. $7\frac{3}{10} =$

Lesson 25: Decompose and compose fractions greater than 1 to express them in various forms.

© 2018 Great Minds®. eureka-math.org

EUREKA MATH

Barbara needed $3\frac{1}{4}$ cups of flour for her recipe. If she measured $\frac{1}{4}$ cup at a time, how many times did she have to fill the measuring cup?

_____

_____

_____

_____

Read   **Draw**   **Write**

Lesson 26:   Compare fractions greater than 1 by reasoning using benchmark
               fractions.

175

© 2018 Great Minds®. eureka-math.org

Name _____ Date 3-14-23

1. a. Plot the following points on the number line without measuring.

   i. $2\frac{7}{8}$ ii. $3\frac{1}{6}$ iii. $\frac{29}{12}$

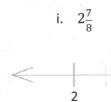

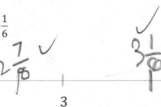

   b. Use the number line in Problem 1(a) to compare the fractions by writing >, <, or =.

   i. $\frac{29}{12}$ ___ $<$ ___ $2\frac{7}{8}$ ii. $\frac{29}{12}$ ___ $<$ ___ $3\frac{1}{6}$

$$12\overline{)29}$$
$$-\ 24 \quad =12\times2$$
$$\overline{5}$$

2. a. Plot the following points on the number line without measuring.

   i. $\frac{70}{9}=\square\frac{\square}{\square}$ ii. $8\frac{2}{4}$ iii. $\frac{25}{3}$

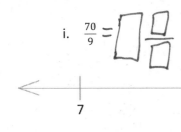

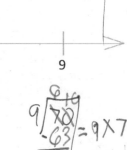

   b. Compare the following by writing >, <, or =.

   i. $8\frac{2}{4}$ ___ $>$ ___ $\frac{25}{3}$ ii. $\frac{70}{9}$ ___ $<$ ___ $8\frac{2}{4}$

$$9\overline{)70} = 9\times7$$
$$-63$$
$$\overline{7} = 9\times1$$

   c. Explain how you plotted the points in Problem 2(a).

   ☆how I Plotted the Points is that I Looked at the nubmers and how close there were to there nubmers and I would Put it there.

EUREKA MATH

Lesson 26:   Compare fractions greater than 1 by reasoning using benchmark fractions.

© 2018 Great Minds®. eureka-math.org

177

3. Compare the fractions given below by writing >, <, or =. Give a brief explanation for each answer, referring to benchmark fractions.

a. $5\frac{1}{3}$ _____ $4\frac{3}{4}$

b. $\frac{12}{6}$ _____ $\frac{25}{12}$

c. $\frac{18}{7}$ _____ $\frac{17}{5}$

d. $5\frac{2}{5}$ _____ $5\frac{5}{8}$

e. $6\frac{2}{3}$ _____ $6\frac{3}{7}$

f. $\frac{31}{7}$ _____ $\frac{32}{8}$

g. $\frac{31}{10}$ _____ $\frac{25}{8}$

h. $\frac{39}{12}$ _____ $\frac{19}{6}$

i. $\frac{49}{50}$ _____ $3\frac{90}{100}$

j. $5\frac{5}{12}$ _____ $5\frac{51}{100}$

Lesson 26: Compare fractions greater than 1 by reasoning using benchmark fractions.

EUREKA MATH

Jeremy ran 27 laps on a track that was $\frac{1}{8}$ mile long. Jimmy ran 15 laps on a track that was $\frac{1}{4}$ mile long. Who ran farther?

_____

_____

_____

_____

Read   **Draw**   **Write**

Lesson 27:   Compare fractions greater than 1 by creating common numerators or denominators.

Name _____   Date 3-15-23

1. Draw a tape diagram to model each comparison. Use >, <, or = to compare.

   a. $3\frac{2}{3}$ __>__ $3\frac{5}{6}$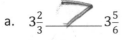

   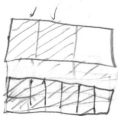 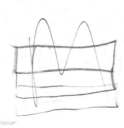

   b. $3\frac{2}{5}$ __<__ $3\frac{6}{10}$

   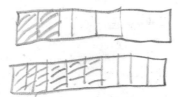

   c. $4\frac{3}{6}$ __>__ $4\frac{1}{3}$

   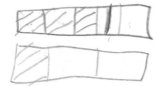

   d. $4\frac{5}{8}$ __<__ $\frac{19}{4}$

2. Use an area model to make like units. Then, use >, <, or = to compare.

   a. $2\frac{3}{5}$ _____ $\frac{18}{7} = 2\frac{4}{7}$

   b. $2\frac{3}{8}$ __>__ $2\frac{1}{3}$

   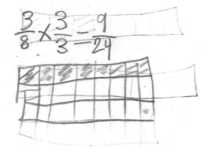

EUREKA MATH

Lesson 27: Compare fractions greater than 1 by creating common numerators or denominators.

183

© 2018 Great Minds®. eureka-math.org

3. Compare each pair of fractions using >, <, or = using any strategy.

a. $5\frac{3}{4}$ _____ $5\frac{3}{8}$

b. $5\frac{2}{5}$ _____ $5\frac{8}{10}$

c. $5\frac{6}{10}$ _____ $\frac{27}{5}$

d. $5\frac{2}{3}$ _____ $5\frac{9}{15}$

e. $\frac{7}{2}$ _____ $\frac{7}{3}$

f. $\frac{12}{3}$ _____ $\frac{15}{4}$

g. $\frac{22}{5}$ _____ $4\frac{2}{7}$

h. $\frac{21}{4}$ _____ $5\frac{2}{5}$

i. $\frac{29}{8}$ _____ $\frac{11}{3}$

j. $3\frac{3}{4}$ _____ $3\frac{4}{7}$

Lesson 27:   Compare fractions greater than 1 by creating common numerators or denominators.

EUREKA
MATH

Name _____   Date 8-16-23

1. The chart to the right shows the distance fourth graders in Ms. Smith's class were able to run before stopping for a rest. Create a line plot to display the data in the table.

Student	Distance (in miles)
Joe	$2\frac{1}{2}$
Arianna	$1\frac{3}{4}$
Bobbi	$2\frac{1}{8}$
Morgan	$1\frac{5}{8}$
Jack	$2\frac{5}{8}$
Saisha	$2\frac{1}{4}$
Tyler	$2\frac{2}{4}$
Jenny	$\frac{5}{8}$
Anson	$2\frac{2}{8}$
Chandra	$2\frac{4}{8}$

Distanca in miles

0   1 $1\frac{3}{4}$ 2 $2\frac{1}{2}$ 3   4

2. Solve each problem.

   a. Who ran a mile farther than Jenny?

   b. Who ran a mile less than Jack?

   c. Two students ran exactly $2\frac{1}{4}$ miles. Identify the students. How many quarter miles did each student run?

   d. What is the difference, in miles, between the longest and shortest distance run?

   e. Compare the distances run by Arianna and Morgan using >, <, or =.

   f. Ms. Smith ran twice as far as Jenny. How far did Ms. Smith run? Write her distance as a mixed number.

   g. Mr. Reynolds ran $1\frac{3}{10}$ miles. Use >, <, or = to compare the distance Mr. Reynolds ran to the distance that Ms. Smith ran. Who ran farther?

3. Using the information in the table and on the line plot, develop and write a question similar to those above. Solve, and then ask your partner to solve. Did you solve in the same way? Did you get the same answer?

Lesson 28:   Solve word problems with line plots.

EUREKA
MATH

Both Allison and Jennifer jogged on Sunday. When asked about their distances, Allison said, "I ran $2\frac{7}{8}$ miles this morning and $3\frac{3}{8}$ miles this afternoon. So, I ran a total of about 6 miles," and Jennifer said, "I ran $3\frac{1}{10}$ miles this morning and $3\frac{3}{10}$ miles this evening. I ran a total of $6\frac{4}{10}$ miles." How do their answers differ?

_____

_____

_____

_____

Read   **Draw**   **Write**

Name _____   Date _3-20-23_

1. Estimate each sum or difference to the nearest half or whole number by rounding. Explain your estimate using words or a number line.

a. $2\frac{1}{12} + 1\frac{7}{8} \approx$ ___4___ ✓

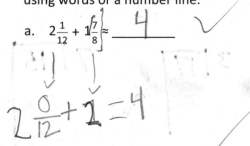

$2\frac{0}{12} + 2 = 4$

b. $1\frac{11}{12} + 5\frac{3}{4} \approx$ ___8___ ✓

$2 + 6 = 8$

c. $8\frac{7}{8} - 2\frac{1}{9} \approx$ ___7___ ✓

$9 - 2 = 7$

d. $6\frac{1}{8} - 2\frac{1}{12} \approx$ ___4___ ✓

$6 - 2 = 4$

e. $3\frac{3}{8} + 5\frac{1}{9} \approx$ ___$8\frac{1}{2}$___ ✓

$3\frac{1}{2} + 5 = 8\frac{1}{2}$

2. Estimate each sum or difference to the nearest half or whole number by rounding. Explain your estimate using words or a number line.

a. $\frac{16}{5} + \frac{11}{4} \approx$ _____ 6

3 + 3

$4 + 4 = 8$

$8 + 4 = 12$

$5\frac{2}{5}$

$3\overline{|17|} = 3 \times 5$
  $\underline{15}$
  2

$1\frac{2}{2}$

$7\overline{|15|} = 7 \times 2$
  $\underline{14}$
  1

b. $\frac{17}{3} - \frac{15}{7} \approx$ _____ 6

$9 - 2 + 1$

$3 + 3 = 6$ $7 + 7 = 14$

$6 + 3 = 9$ $14 + 7 = 21$

c. $\frac{59}{10} + \frac{26}{10} \approx$ _____ 10

6   4

$10 + 10 = 20$

$10 + 20 = 30$

3. Montoya's estimate for $8\frac{5}{8} - 2\frac{1}{3}$ was 7. Julio's estimate was $6\frac{1}{2}$. Whose estimate do you think is closer to the actual difference? Explain.

Julio's estimate was $6\frac{1}{2}$ the closer to the actual difference.

$8\frac{5}{8} - 2\frac{1}{3} = 6\frac{1}{2}$

$8\frac{1}{2} - 2\frac{0}{3} = 6\frac{1}{2}$

4. Use benchmark numbers or mental math to estimate the sum or difference.

a. $14\frac{3}{4} + 29\frac{11}{12} = 45$ ✓ $15 + 30 = 45$	b. $3\frac{5}{12} + 54\frac{5}{8} = 58$ ✓ $3\frac{1}{2} + 54\frac{1}{2} = 58\frac{1}{2}$
c. $17\frac{4}{5} - 8\frac{7}{12} = 9\frac{1}{2}$ ✓ $18 - 8\frac{1}{2} = 9\frac{1}{2}$	d. $\frac{65}{8} - \frac{37}{6} = 2$ $7 - 5 = 2$

EUREKA MATH

One board measures 2 meters 70 centimeters. Another measures 87 centimeters. What is the total length of the two boards expressed in meters and centimeters?

_____

_____

_____

_____

Read   **Draw**   **Write**

Name _____   Date _____

1. Solve.

   a. $3\frac{1}{4} + \frac{1}{4}$

   b. $7\frac{3}{4} + \frac{1}{4}$

   c. $\frac{3}{8} + 5\frac{2}{8}$

   d. $\frac{1}{8} + 6\frac{7}{8}$

2. Complete the number sentences.

a. $4\frac{7}{8} +$ _____ $= 5$	b. $7\frac{2}{5} +$ _____ $= 8$
c. $3 = 2\frac{1}{6} +$ _____	d. $12 = 11\frac{1}{12} +$ _____

3. Use a number bond and the arrow way to show how to make one. Solve.

   a. $2\frac{3}{4} + \frac{2}{4}$

   b. $3\frac{3}{5} + \frac{3}{5}$

4. Solve.

a. $4\frac{2}{3} + \frac{2}{3}$	b. $3\frac{3}{5} + \frac{4}{5}$
c. $5\frac{4}{6} + \frac{5}{6}$	d. $\frac{7}{8} + 6\frac{4}{8}$
e. $\frac{7}{10} + 7\frac{9}{10}$	f. $9\frac{7}{12} + \frac{11}{12}$
g. $2\frac{70}{100} + \frac{87}{100}$	h. $\frac{50}{100} + 16\frac{78}{100}$

Lesson 30:   Add a mixed number and a fraction.

EUREKA
MATH®

5. To solve $7\frac{9}{10} + \frac{5}{10}$, Maria thought, "$7\frac{9}{10} + \frac{1}{10} = 8$ and $8 + \frac{4}{10} = 8\frac{4}{10}$."

   Paul thought, "$7\frac{9}{10} + \frac{5}{10} = 7\frac{14}{10} = 7 + \frac{10}{10} + \frac{4}{10} = 8\frac{4}{10}$." Explain why Maria and Paul are both right.

Name _____   Date _____

Solve.

1. $3\frac{2}{5} + $ _____ $= 4$

2. $2\frac{3}{8} + \frac{7}{8}$

Marta has 2 meters 80 centimeters of cotton cloth and 3 meters 87 centimeters of linen cloth. What is the total length of both pieces of cloth?

_____

_____

_____

_____

Read   **Draw**   **Write**

Name _____   Date _____

1. Solve.

   a. $3\frac{1}{3}$ + $2\frac{2}{3}$ = 5 + $\frac{3}{3}$ =

   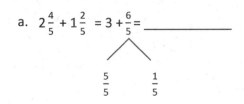

   b. $4\frac{1}{4} + 3\frac{2}{4}$

   c. $2\frac{2}{6} + 6\frac{4}{6}$

2. Solve. Use a number line to show your work.

   a. $2\frac{4}{5} + 1\frac{2}{5}$ = 3 + $\frac{6}{5}$ = _____

   b. $1\frac{3}{4} + 3\frac{3}{4}$

   c. $3\frac{3}{8} + 2\frac{6}{8}$

3. Solve. Use the arrow way to show how to make one.

   a. $2\frac{4}{6} + 1\frac{5}{6} = 3\frac{4}{6} + \frac{5}{6} =$

   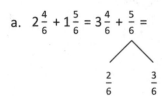

   b. $1\frac{3}{4} + 3\frac{3}{4}$

   c. $3\frac{3}{8} + 2\frac{6}{8}$

4. Solve. Use whichever method you prefer.

   a. $1\frac{3}{5} + 3\frac{4}{5}$

   b. $2\frac{6}{8} + 3\frac{7}{8}$

   c. $3\frac{8}{12} + 2\frac{7}{12}$

Name _____   Date _____

Solve.

1. $2\frac{3}{8} + 1\frac{5}{8}$

2. $3\frac{4}{5} + 2\frac{3}{5}$

Meredith had 2 m 65 cm of ribbon. She used 87 cm of the ribbon. How much ribbon did she have left?

_____

_____

_____

_____

Read   **Draw**   **Write**

Name _____   Date 3-22-23

1. Subtract. Model with a number line or the arrow way.

    a. $3\frac{3}{4} - \frac{1}{4} =$ $3\frac{2}{4}$ ✓

    $2 + \frac{4}{4}$

    b. $4\frac{7}{10} - \frac{3}{10} =$ 4 $\frac{4}{10}$ ✓

    $4\frac{10}{10}$

    c. $5\frac{1}{3} - \frac{2}{3} =$ 4 $\frac{2}{3}$ ✓

    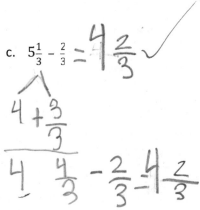

    $4 + \frac{3}{3}$

    $4 \quad \frac{4}{3} - \frac{2}{3} = 4\frac{2}{3}$

    d. $9\frac{3}{5} - \frac{4}{5} =$ 8 $\frac{4}{5}$

    $8 + \frac{5}{5}$

    $8 \quad \frac{8}{5} - \frac{4}{5} = 8\frac{4}{5}$

2. Use decomposition to subtract the fractions. Model with a number line or the arrow way.

    a. $5\frac{3}{5} - \frac{4}{5}$

    $\overbrace{\quad}$
    $\frac{3}{5} \quad \frac{1}{5}$

    b. $4\frac{1}{4} - \frac{2}{4} =$ 3 $\frac{3}{4}$ ✓

    $3 + \frac{4}{4}$

    $3\frac{5}{4} + \frac{2}{4} = 4\frac{3}{4}$

    c. $5\frac{1}{3} - \frac{2}{3} =$ 4 $\frac{2}{3}$ ✓

    $4 + \frac{3}{3}$

    $4\frac{4}{3} - \frac{2}{3} = 5\frac{2}{3}$

    d. $2\frac{3}{8} - \frac{5}{8} =$ 1 $\frac{6}{8}$ ✓

    $1 + \frac{8}{8}$

    $1\frac{11}{8} - \frac{5}{8} = 1\frac{6}{8}$

3. Decompose the total to subtract the fractions.

a. $3\frac{1}{8} - \frac{3}{8} = 2\frac{1}{8} + \frac{5}{8} = 2\frac{6}{8}$

$2\frac{1}{8}$ ⁀ 1

b. $5\frac{1}{8} - \frac{7}{8} = 4\frac{2}{8}$ ✓

$4 + \frac{8}{8}$

$4\frac{9}{8} - \frac{7}{8} = 4\frac{2}{8}$

c. $5\frac{3}{5} - \frac{4}{5} = 4\frac{4}{5}$

$4 + \frac{5}{5}$

$4\frac{8}{5} - \frac{4}{5} = 4\frac{4}{5}$

d. $5\frac{4}{6} - \frac{5}{6} = 4\frac{5}{6}$ ✓

$4 + \frac{6}{6}$

$4\frac{10}{6} - \frac{5}{6} = 4\frac{5}{6}$

e. $6\frac{4}{12} - \frac{7}{12} = 5\frac{9}{12}$ ✓

$5 + \frac{12}{12}$

$5\frac{16}{12} - \frac{7}{12} = 5\frac{9}{12}$

f. $9\frac{1}{8} - \frac{5}{8} = 8\frac{4}{8}$ ✓

$8 + \frac{8}{8}$

$8\frac{9}{8} - \frac{5}{8} = 8\frac{4}{8}$

g. $7\frac{1}{6} - \frac{5}{6} = 6\frac{2}{6}$ ✓

$6 + \frac{6}{6}$

$6\frac{7}{6} - \frac{5}{6} = 6\frac{2}{6}$

h. $8\frac{3}{10} - \frac{4}{10} = 7\frac{9}{10}$

$7 + \frac{10}{10}$

$7\frac{13}{10} - \frac{4}{10} = 7\frac{9}{10}$

i. $12\frac{3}{5} - \frac{4}{5} = 11\frac{4}{5}$ ✓

$11 + \frac{5}{5}$

$11\frac{8}{5} - \frac{4}{5} = 11\frac{4}{5}$

j. $11\frac{2}{6} - \frac{5}{6} = 10\frac{3}{6}$ ✓

$10 + \frac{6}{6}$

$10\frac{8}{6} - \frac{5}{6} = 10\frac{3}{6}$

Lesson 32: Subtract a fraction from a mixed number.

EUREKA MATH®

Jeannie's pumpkin had a weight of 3 kg 250 g in August and 4 kg 125 g in October. What was the difference in weight from August to October?

_____

_____

_____

_____

Read   Draw   Write

Name _____   Date _____

1. Write a related addition sentence. Subtract by counting on. Use a number line or the arrow way to help.
    The first one has been partially done for you.

    a. $3\frac{1}{3} - 1\frac{2}{3} =$ _____

        $1\frac{2}{3} +$ _____ $= 3\frac{1}{3}$

    b. $5\frac{1}{4} - 2\frac{3}{4} =$ _____

2. Subtract, as shown in Problem 2(a), by decomposing the fractional part of the number you are
    subtracting. Use a number line or the arrow way to help you.

    a. $3\frac{1}{4} - 1\frac{3}{4} = 2\frac{1}{4} - \frac{3}{4} = 1\frac{2}{4}$

        $\frac{1}{4} \quad \frac{2}{4}$

    b. $4\frac{1}{5} - 2\frac{4}{5}$

    c. $5\frac{3}{7} - 3\frac{6}{7}$

3. Subtract, as shown in Problem 3(a), by decomposing to take one out.

   a. $5\frac{3}{5} - 2\frac{4}{5} = 3\frac{3}{5} - \frac{4}{5}$

       $2\frac{3}{5}$ 1

   b. $4\frac{3}{6} - 3\frac{5}{6}$

   c. $8\frac{3}{10} - 2\frac{7}{10}$

4. Solve using any method.

   a. $6\frac{1}{4} - 3\frac{3}{4}$ b. $5\frac{1}{8} - 2\frac{7}{8}$

   c. $8\frac{3}{12} - 3\frac{8}{12}$ d. $5\frac{1}{100} - 2\frac{97}{100}$

Lesson 33:   Subtract a mixed number from a mixed number.

EUREKA
MATH®

Name _____   Date _____

Solve using any strategy.

1. $4\frac{2}{3} - 2\frac{1}{3}$

2. $12\frac{5}{8} - 8\frac{7}{8}$

There were $4\frac{1}{8}$ pizzas. Benny took $\frac{2}{8}$ of a pizza. How many pizzas are left?

_____

_____

_____

_____

Read   **Draw**   **Write**

Lesson 34:   Subtract mixed numbers.   **223**

Name _____ Date 3-23-23

1. Subtract.

a. $4\frac{1}{3} - \frac{2}{3}$

3 ╱ ╲ $\frac{4}{3}$

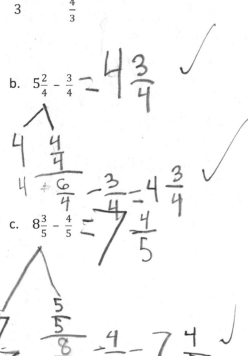

b. $5\frac{2}{4} - \frac{3}{4} = 4\frac{3}{4}$ ✓

4 ╱ ╲ $\frac{4}{4}$

$4 \neq \frac{6}{4} - \frac{3}{4} = 4\frac{3}{4}$ ✓

c. $8\frac{3}{5} - \frac{4}{5} = 7\frac{4}{5}$ ✓

7 ╱ ╲ $\frac{5}{5}$

$7 \frac{8}{5} - \frac{4}{5} = 7\frac{4}{5}$ ✓

2. Subtract the ones first.

a. $3\frac{1}{4} - 1\frac{3}{4} = 2\frac{1}{4} - \frac{3}{4} = 1\frac{2}{4}$

1 ╱ ╲ $\frac{5}{4}$

b. $4\frac{2}{5} - 1\frac{3}{5} = 2\frac{4}{5}$

3 ╱ ╲ $\frac{5}{5}$

$3\frac{7}{5} - 1\frac{3}{5} = 2\frac{4}{5}$ ✓

c. $5\frac{2}{6} - 3\frac{5}{6} =$ **1$\frac{3}{6}$** ✓

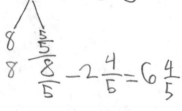

$4\frac{6}{6}$

$4\frac{8}{6} - 3\frac{5}{6} = 1\frac{3}{6}$

d. $9\frac{3}{5} - 2\frac{4}{5} =$ **6$\frac{4}{5}$**

$8\frac{5}{5}$

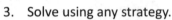

$8\frac{8}{5} - 2\frac{4}{5} = 6\frac{4}{5}$ ✓

3. Solve using any strategy.

a. $7\frac{3}{8} - 2\frac{5}{8} =$ **4$\frac{6}{8}$** ✓

$6\frac{8}{8}$

$6\frac{11}{8} - 2\frac{5}{8} = 4\frac{6}{8}$

b. $6\frac{4}{10} - 3\frac{8}{10} =$ **2$\frac{6}{10}$** ✓

$5\frac{10}{10}$

$5\frac{14}{10} - 3\frac{8}{10} = 2\frac{6}{10}$

c. $8\frac{3}{12} - 3\frac{8}{12} =$ **4$\frac{7}{12}$**

$7\frac{12}{12}$

$7\frac{15}{12} - 3\frac{8}{12} = 4\frac{7}{12}$

d. $14\frac{2}{50} - 6\frac{43}{50} =$ **7$\frac{9}{50}$**

$13\frac{56}{50}$

$13\frac{52}{50} - 6\frac{43}{50} = 7\frac{9}{50}$ ✓

Lesson 34:   Subtract mixed numbers.

© 2018 Great Minds®. eureka-math.org

$\begin{array}{r}{}^{4}12\\ 5\!\!\!\!/2\\ -43\\ \hline 9\end{array}$

EUREKA MATH®

Mary Beth is knitting scarves that are 1 meter long. If she knits 54 centimeters of a scarf each night for 3 nights, how many scarves will she complete? How much more does she need to knit to complete another scarf?

_____

_____

_____

_____

Read   **Draw**   **Write**

Lesson 35:   Represent the multiplication of *n* times *a/b* as (*n* × *a*)/*b* using the
                associative property and visual models.

229

© 2018 Great Minds®. eureka-math.org

Name _____   Date 3-24-23

1. Draw and label a tape diagram to show the following are true.

   a.   8 fifths = 4 × (2 fifths) = (4 × 2) fifths

   b.   10 sixths = 5 × (2 sixths) = (5 × 2) sixths

2. Write the expression in unit form to solve.

   a. $7 \times \frac{2}{3} = \frac{7 \times 2}{3} = \frac{14}{3} = 4 \cdot \frac{2}{3}$ ✓

   $\frac{2}{8} + 7 = \frac{9}{3}$

   b. $4 \times \frac{2}{4} = \frac{4 \times 2}{4} = \frac{8}{4} = 2$

   c. $16 \times \frac{3}{8} = \frac{16 \times 3}{8} = \frac{48}{8} = 6$

   $\frac{3}{8} + \frac{3}{8} + \frac{3}{8} \cdots$

   d. $6 \times \frac{5}{8} = \frac{6 \times 5}{8} = \frac{30}{8} = 3 \frac{6}{8}$

EUREKA
MATH

Lesson 35:   Represent the multiplication of n times a/b as (n × a)/b using the
associative property and visual models.

© 2018 Great Minds®. eureka-math.org

231

3. Solve.

a. $7 \times \frac{4}{9} = \frac{7 \times 4}{9} = \frac{28}{9} = 3\frac{1}{9}$ ✓

b. $6 \times \frac{3}{5} = \frac{6 \times 3}{5} = \frac{18}{5} = 3\frac{3}{5}$ ✓

c. $8 \times \frac{3}{4} \quad \frac{8 \times 3}{4} = \frac{24}{4} = 6$ ✓

d. $16 \times \frac{3}{8} = 6$

e. $12 \times \frac{7}{10} \quad \frac{12 \times 7}{10} = \frac{84}{10} = 8\frac{4}{10}$ ✓

f. $3 \times \frac{54}{100}$

4. Maria needs $\frac{3}{5}$ yard of fabric for each costume. How many yards of fabric does she need for 6 costumes?

$6 \times \frac{3}{5} = \frac{6 \times 3}{5} = \frac{18}{5} = 3\frac{3}{5}$

maria needs $3\frac{3}{5}$ for 6 costumes

Lesson 35:   Represent the multiplication of n times a/b as (n × a)/b using the associative property and visual models.

© 2018 Great Minds®. eureka-math.org

EUREKA MATH

Rhonda exercised for $\frac{5}{6}$ hour every day for 5 days. How many total hours did Rhonda exercise?

Read   **Draw**   **Write**

Lesson 36:   Represent the multiplication of *n* times *a/b* as (*n* × *a*)/*b* using the
associative property and visual models.

235

© 2018 Great Minds®. eureka-math.org

Name _____   Date _____

1. Draw a tape diagram to represent

    $\frac{3}{4}+\frac{3}{4}+\frac{3}{4}+\frac{3}{4}$.

2. Draw a tape diagram to represent

    $\frac{7}{12}+\frac{7}{12}+\frac{7}{12}$.

Write a multiplication expression equal to

$\frac{3}{4}+\frac{3}{4}+\frac{3}{4}+\frac{3}{4}$.

Write a multiplication expression equal to

$\frac{7}{12}+\frac{7}{12}+\frac{7}{12}$.

3. Rewrite each repeated addition problem as a multiplication problem and solve. Express the result as a mixed number. The first one has been started for you.

    a. $\frac{7}{5}+\frac{7}{5}+\frac{7}{5}+\frac{7}{5}=4\times\frac{7}{5}=\frac{4\times7}{5}=$

    b. $\frac{9}{10}+\frac{9}{10}+\frac{9}{10}$

    c. $\frac{11}{12}+\frac{11}{12}+\frac{11}{12}+\frac{11}{12}+\frac{11}{12}$

Lesson 36:   Represent the multiplication of *n* times *a/b* as (*n* × *a*)/*b* using the associative property and visual models.

© 2018 Great Minds®. eureka-math.org

237

4. Solve using any method. Express your answers as whole or mixed numbers.

   a. $8 \times \frac{2}{3}$

   b. $12 \times \frac{3}{4}$

   c. $50 \times \frac{4}{5}$

   d. $26 \times \frac{7}{8}$

5. Morgan poured $\frac{9}{10}$ liter of punch into each of 6 bottles. How many liters of punch did she pour in all?

6. A recipe calls for $\frac{3}{4}$ cup rice. How many cups of rice are needed to make the recipe 14 times?

7. A butcher prepared 120 sausages using $\frac{3}{8}$ pound of meat for each. How many pounds did he use in all?

Lesson 36: Represent the multiplication of n times a/b as (n × a)/b using the associative property and visual models.

EUREKA MATH®

Name _____   Date _____

Solve using any method.

1. $7 \times \frac{3}{4}$

2. $9 \times \frac{2}{5}$

3. $60 \times \frac{5}{8}$

Lesson 36:   Represent the multiplication of *n* times *a/b* as (*n* × *a*)/*b* using the
associative property and visual models.

239

© 2018 Great Minds®. eureka-math.org

The baker needs $\frac{5}{8}$ cup of raisins to make 1 batch of cookies. How many cups of raisins does he need to make 7 batches of cookies?

_____

_____

_____

_____

Read   Draw   Write

Lesson 37:   Find the product of a whole number and a mixed number using the
              distributive property.

© 2018 Great Minds®. eureka-math.org

241

Name _____   Date _____

1. Draw tape diagrams to show two ways to represent 2 units of $4\frac{2}{3}$.

Write a multiplication expression to match each tape diagram.

2. Solve the following using the distributive property. The first one has been done for you. (As soon as you are ready, you may omit the step that is in line 2.)

a. $\quad 3 \times 6\frac{4}{5} = 3 \times \left(6 + \frac{4}{5}\right)$ $\qquad = (3 \times 6) + \left(3 \times \frac{4}{5}\right)$ $\qquad = 18 + \frac{12}{5}$ $\qquad = 18 + 2\frac{2}{5}$ $\qquad = 20\frac{2}{5}$	b. $\quad 2 \times 4\frac{2}{3}$
c. $\quad 3 \times 2\frac{5}{8}$	d. $\quad 2 \times 4\frac{7}{10}$

EUREKA MATH   **Lesson 37:**   Find the product of a whole number and a mixed number using the distributive property.   **243**

© 2018 Great Minds®. eureka-math.org

e. $3 \times 7\frac{3}{4}$	f. $6 \times 3\frac{1}{2}$
g. $4 \times 9\frac{1}{5}$	h. $5\frac{6}{8} \times 4$

3. For one dance costume, Saisha needs $4\frac{2}{3}$ feet of ribbon. How much ribbon does she need for 5 identical costumes?

Lesson 37:   Find the product of a whole number and a mixed number using the
                      distributive property.

EUREKA MATH

Name _____   Date _____

Multiply. Write each product as a mixed number.

1. $4 \times 5\frac{3}{8}$

2. $4\frac{3}{10} \times 3$

Lesson 37:   Find the product of a whole number and a mixed number using the
              distributive property.

© 2018 Great Minds®. eureka-math.org

245

Eight students are on a relay team. Each runs $1\frac{3}{4}$ kilometers. How many total kilometers does their team run?

_____

_____

_____

_____

Read   **Draw**   **Write**

Lesson 38:   Find the product of a whole number and a mixed number using the
distributive property.

© 2018 Great Minds®. eureka-math.org

247

Name _____ Date _____

1. Fill in the unknown factors.

   a. $7 \times 3\frac{4}{5} = (\underline{} \times 3) + (\underline{} \times \frac{4}{5})$

   b. $3 \times 12\frac{7}{8} = (3 \times \underline{}) + (3 \times \underline{})$

2. Multiply. Use the distributive property.

   a. $7 \times 8\frac{2}{5}$

   b. $4\frac{5}{6} \times 9$

   c. $3 \times 8\frac{11}{12}$

   d. $5 \times 20\frac{8}{10}$

Lesson 38:   Find the product of a whole number and a mixed number using the
              distributive property.

e. $25\dfrac{4}{100} \times 4$

3. The distance around the park is $2\dfrac{5}{10}$ miles. Cecilia ran around the park 3 times. How far did she run?

4. Windsor the dog ate $4\dfrac{3}{4}$ snack bones each day for a week. How many bones did Windsor eat that week?

Lesson 38:   Find the product of a whole number and a mixed number using the distributive property.

EUREKA MATH

Name _____   Date _____

1. Fill in the unknown factors.

$$8 \times 5\frac{2}{3} = (\underline{} \times 5) + (\underline{} \times \frac{2}{3})$$

2. Multiply. Use the distributive property.

$$6\frac{5}{8} \times 7$$

EUREKA
MATH

Lesson 38:   Find the product of a whole number and a mixed number using the
distributive property.

251

© 2018 Great Minds®. eureka-math.org

Name _____ Date _____

Use the RDW process to solve.

1. Tameka ran $2\frac{5}{8}$ miles. Her sister ran twice as far. How far did Tameka's sister run?

2. Natasha's sculpture was $5\frac{3}{16}$ inches tall. Maya's was 4 times as tall. How much shorter was Natasha's sculpture than Maya's?

3. A seamstress needs $1\frac{5}{8}$ yards of fabric to make a child's dress. She needs 3 times as much fabric to make a woman's dress. How many yards of fabric does she need for both dresses?

4. A piece of blue yarn is $5\frac{2}{3}$ yards long. A piece of pink yarn is 5 times as long as the blue yarn. Bailey tied them together with a knot that used $\frac{1}{3}$ yard from each piece of yarn. What is the total length of the yarn tied together?

5. A truck driver drove $35\frac{2}{10}$ miles before he stopped for breakfast. He then drove 5 times as far before he stopped for lunch. How far did he drive that day before his lunch break?

6. Mr. Washington's motorcycle needs $5\frac{5}{10}$ gallons of gas to fill the tank. His van needs 5 times as much gas to fill it. If Mr. Washington pays \$3 per gallon for gas, how much will it cost him to fill both the motorcycle and the van?

EUREKA
MATH

Name _____   Date _____

Use the RDW process to solve.

Jeff has ten packages that he wants to mail. Nine identical packages weigh $2\frac{7}{8}$ pounds each. A tenth package weighs two times as much as one of the other packages. How many pounds do all ten packages weigh?

Name _____ Date _____

1. The chart to the right shows the height of some football players.

   a. Use the data to create a line plot at the bottom of this page and to answer the questions below.

   b. What is the difference in height of the tallest and shortest players?

   c. Player I and Player B have a combined height that is $1\frac{1}{8}$ feet taller than a school bus. What is the height of a school bus?

Player	Height (in feet)
A	$6\frac{1}{4}$
B	$5\frac{7}{8}$
C	$6\frac{1}{2}$
D	$6\frac{1}{4}$
E	$6\frac{2}{8}$
F	$5\frac{7}{8}$
G	$6\frac{1}{8}$
H	$6\frac{5}{8}$
I	$5\frac{6}{8}$
J	$6\frac{1}{8}$

Lesson 40: Solve word problems involving the multiplication of a whole number and a fraction including those involving line plots.

© 2018 Great Minds®. eureka-math.org

257

2. One of the players on the team is now 4 times as tall as he was at birth, when he measured $1\frac{5}{8}$ feet. Who is the player?

3. Six of the players on the team weigh over 300 pounds. Doctors recommend that players of this weight drink at least $3\frac{3}{4}$ quarts of water each day. At least how much water should be consumed per day by all 6 players?

4. Nine of the players on the team weigh about 200 pounds. Doctors recommend that people of this weight each eat about $3\frac{7}{10}$ grams of carbohydrates per pound each day. About how many combined grams of carbohydrates should these 9 players eat per pound each day?

Lesson 40:   Solve word problems involving the multiplication of a whole number and a fraction including those involving line plots.

EUREKA
MATH

Name _____   Date _____

Coach Taylor asked his team to record the distance they ran during practice.

The distances are listed in the table.

1. Use the table to locate the incorrect data on the line plot.

   Circle any incorrect points.

   Mark any missing points.

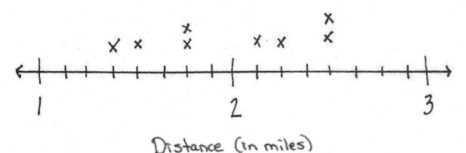

2. Of the team members who ran $1\frac{6}{8}$ miles, how many miles did those team members run combined?

Team Members	Distance (in miles)
Alec	$1\frac{3}{4}$
Henry	$1\frac{1}{2}$
Charles	$2\frac{1}{8}$
Steve	$1\frac{3}{4}$
Pitch	$2\frac{2}{4}$
Raj	$1\frac{6}{8}$
Pam	$2\frac{1}{2}$
Tony	$1\frac{3}{8}$

EUREKA
MATH

Lesson 40:   Solve word problems involving the multiplication of a whole number
and a fraction including those involving line plots.

259

© 2018 Great Minds®. eureka-math.org

Jackie's paper chain was 5 times as long as Sammy's, which measured $2\frac{75}{100}$ meters. What was the total length of both their chains?

_____

_____

_____

_____

Read   **Draw**   **Write**

Lesson 41:   Find and use a pattern to calculate the sum of all fractional parts
              between 0 and 1. Share and critique peer strategies.

261

© 2018 Great Minds®. eureka-math.org

Name _____   Date _____

1. Find the sums.

   a. $\dfrac{0}{3} + \dfrac{1}{3} + \dfrac{2}{3} + \dfrac{3}{3}$

   b. $\dfrac{0}{4} + \dfrac{1}{4} + \dfrac{2}{4} + \dfrac{3}{4} + \dfrac{4}{4}$

   c. $\dfrac{0}{5} + \dfrac{1}{5} + \dfrac{2}{5} + \dfrac{3}{5} + \dfrac{4}{5} + \dfrac{5}{5}$

   d. $\dfrac{0}{6} + \dfrac{1}{6} + \dfrac{2}{6} + \dfrac{3}{6} + \dfrac{4}{6} + \dfrac{5}{6} + \dfrac{6}{6}$

   e. $\dfrac{0}{7} + \dfrac{1}{7} + \dfrac{2}{7} + \dfrac{3}{7} + \dfrac{4}{7} + \dfrac{5}{7} + \dfrac{6}{7} + \dfrac{7}{7}$

   f. $\dfrac{0}{8} + \dfrac{1}{8} + \dfrac{2}{8} + \dfrac{3}{8} + \dfrac{4}{8} + \dfrac{5}{8} + \dfrac{6}{8} + \dfrac{7}{8} + \dfrac{8}{8}$

2. Describe a pattern you notice when adding the sums of fractions with even denominators as opposed to those with odd denominators.

3. How would the sums change if the addition started with the unit fraction rather than with 0?

EUREKA MATH®

Lesson 41:   Find and use a pattern to calculate the sum of all fractional parts
              between 0 and 1. Share and critique peer strategies.

© 2018 Great Minds®. eureka-math.org

263

4. Find the sums.

a. $\frac{0}{10} + \frac{1}{10} + \frac{2}{10} + \cdots + \frac{10}{10}$

b. $\frac{0}{12} + \frac{1}{12} + \frac{2}{12} + \cdots + \frac{12}{12}$

c. $\frac{0}{15} + \frac{1}{15} + \frac{2}{15} + \cdots + \frac{15}{15}$

d. $\frac{0}{25} + \frac{1}{25} + \frac{2}{25} + \cdots + \frac{25}{25}$

e. $\frac{0}{50} + \frac{1}{50} + \frac{2}{50} + \cdots + \frac{50}{50}$

f. $\frac{0}{100} + \frac{1}{100} + \frac{2}{100} + \cdots + \frac{100}{100}$

5. Compare your strategy for finding the sums in Problems 4(d), 4(e), and 4(f) with a partner.

6. How can you apply this strategy to find the sum of all the whole numbers from 0 to 100?

Lesson 41:   Find and use a pattern to calculate the sum of all fractional parts
             between 0 and 1. Share and critique peer strategies.

EUREKA
MATH

Name _____   Date _____

Find the sums.

1. $\dfrac{0}{20} + \dfrac{1}{20} + \dfrac{2}{20} + \cdots + \dfrac{20}{20}$

2. $\dfrac{0}{200} + \dfrac{1}{200} + \dfrac{2}{200} + \cdots + \dfrac{200}{200}$

Lesson 41:   Find and use a pattern to calculate the sum of all fractional parts between 0 and 1. Share and critique peer strategies.

© 2018 Great Minds®. eureka-math.org

265

Credits

Great Minds® has made every effort to obtain permission for the reprinting of all copyrighted material. If any owner of copyrighted material is not acknowledged herein, please contact Great Minds for proper acknowledgment in all future editions and reprints of this module.